HELLO!
HELLO!
A
B
Stand on mountain "A" and shout.
Your voice will bounce off mountain "B"
and come back as an echo.
Radar uses echoes, too.
(See page 90.)

Try blowing across the top of a pop bottle.
The column of air inside will vibrate and sound sort of like a horn.
(See page 70.)
Put a blade of grass between your thumbs and blow.
The sound will be like a woodwind with a reed—
a clarinet, for instance.
(See page 71.)

Throw a stone into the water.
PLOP
Waves will spread out
in larger and larger circles,
just like radio waves.
(See page 74.)

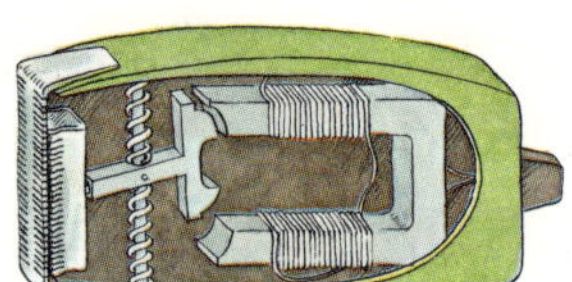
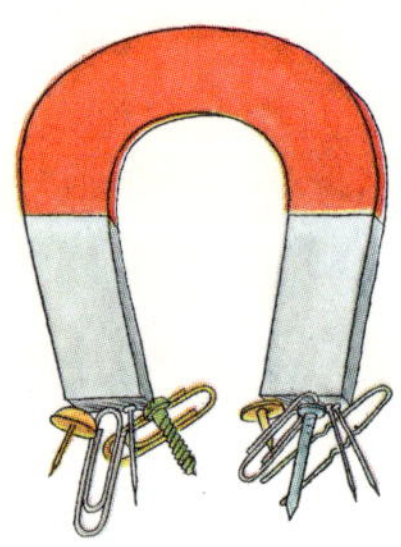
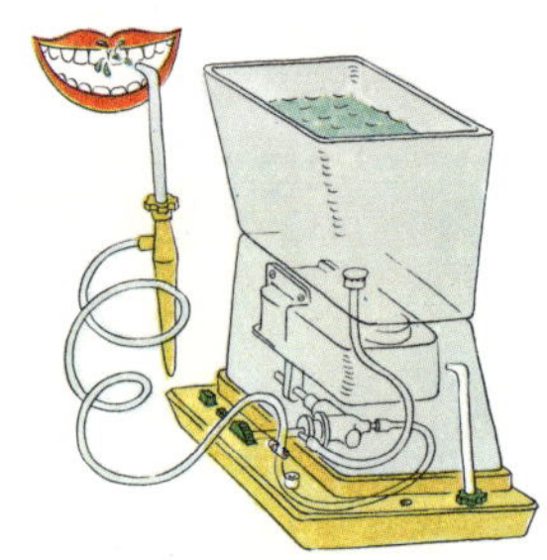

Joe Kaufman's

WHAT MAKES IT GO?

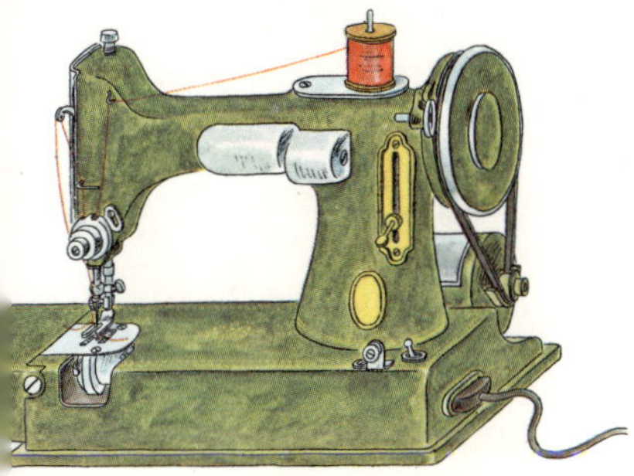

WHAT MAKES IT WORK?

WHAT MAKES IT FLY?

WHAT MAKES IT FLOAT?

Written and Illustrated by

JOE KAUFMAN

Golden Press · New York

Western Publishing Company, Inc.
Racine, Wisconsin

Library of Congress Catalog Card Number: 70-154824

CONTENTS

A NOTE TO PARENTS: This book was designed to help answer one of the most frequently asked questions of childhood—"What makes it work?" It is a basic introduction to the mechanics, and the concepts behind the mechanics, of a variety of appliances, machines, and vehicles. Although it was designed primarily for young readers from six to ten, older children and adults will surely find much to learn and enjoy. In fact, the entire family can share many exciting and rewarding hours studying the great ideas of inventors of the past as they are incorporated in the things we use today.

We hope the book will awaken the child's interest in the roles science and technology play in giving us the things we take for granted, and perhaps encourage him to go on to more detailed works in areas of particular interest. It may also be a beginning for those who will one day make the discoveries that lie ahead.

1.

Early man, many thousands of years ago, found that getting along was pretty rough. He had only his bare hands to use for hunting, fishing, and digging up roots to eat. Where could he sleep? He had only a cave to crawl into at night, with maybe a rock for a pillow. What could he do about cold and rain and windstorm? And let's not forget those howling jungle beasts.

So what did man do? He had a brain that could think and hands that could make things. So he slowly began to invent ways to make his life safer and more pleasant. He made tools for hunting, fishing, and building. At first, the tools were just sharpened stones, but they were a good start. He discovered that fire was good for keeping warm and that water was good for washing. With fur from wild animals, he invented clothes.

And one invention led to another. Perhaps he discovered that a cross-section of a tree trunk would roll. "What can I do with a rolling wooden wheel?" he asked. Soon he thought of forming other logs into a box, attaching his wheels, and there it was — a wagon. "Now, which animal can I train to pull my invention?" he asked, staring hard at a horse. Of course, one man didn't do all this inventing. It took many men, thinking and inventing over a long period of time, to come up with that first wagon.

2.

3.

And men continued to think up all kinds of things, until now we have ships, trains, autos, and planes — stoves, locks, pianos, and telephones. And now we have men exploring space in rockets. Inventions have brought man a long way since that first wheel.

Some modern inventions are simple and some are very complicated. How the simple ones work is easy to understand, but the complicated ones are tough. This book of explanations should help. Good luck!

4.

A TRICYCLE has three wheels – a big one in front and two small ones behind. The front wheel has two pedals attached to it. When you push them around with your feet, the pedals turn the wheel and the tricycle goes. The two rear wheels follow along and keep the tricycle balanced. You steer with handlebars attached to the front wheel. To go backwards, you can pedal backwards. A tricycle doesn't have any brakes, but you can stop it by stopping the pedals with your feet. The pedals stop the front wheel, and the tricycle stops, too.

A BICYCLE only has two wheels, and it takes a little time to learn to balance yourself on them. The pedals drive the back wheel, not the front, and to do this they are attached to a big sprocket wheel (a wheel with teeth). A loop of chain runs around the sprocket wheel and a gear (a smaller toothed wheel) attached to the rear axle. The links of the chain fit over the teeth, so that when you push the pedals, the sprocket wheel turns the chain, the chain turns the gear and the rear wheel, and the bike is pushed ahead.

There are different gears for starting off, climbing hills, and cruising along. The size of the gear determines the number of times the rear wheel goes around when you push the pedals around once. The more turns the rear wheel makes in one cycle of the pedals, the faster you go . . . but the harder it is to push the pedals!

A MOTORCYCLE works very much like a bicycle, but it is driven by a small automobile-type engine instead of leg-power. The engine drives the rear wheel, sometimes with a chain (just like a bicycle) and sometimes with a driveshaft (like a car). The handlebars steer the motorcycle, and the right handgrip is the throttle. By twisting it, the driver controls the flow of fuel to the engine and so controls the speed.

To slow down or stop there are two brakes, a hand brake for the front wheel and a foot brake for the rear. Both cause a curved piece of metal inside the wheel to press out against the brake drum, stopping it and the wheel from turning. The system of gears, or transmission, is very much like an automobile transmission, which you can read about on page 14.

Because a motorcycle runs much faster than a bicycle, it needs shock absorbers to smooth out the ride. Otherwise, it would go out of control if it hit a bad bump. The shock absorbers are heavy-duty springs, attached to the wheel supports.

When AUTOMOBILES were invented to replace carriages pulled by horses, they were made to look

Automobiles didn't always run by gasoline. These three used electric batteries for their power.

Automobiles that used gasoline engines for their power soon became the most popular. These are

very much like those carriages. Sometimes a make-believe horse head would be fastened to the front.

These three used steam engines which were smaller versions of the powerful railroad steam engines.

early gasoline cars. Some people think we should go back to electric cars, which were much cleaner.

The AUTOMOBILE is a complicated piece of machinery, but it is carefully designed so that most people can run it. To start it, the driver turns the ignition key, sending electricity from the battery to run the little starting motor which gets the big engine going. Gas starts feeding to the engine. Then everything begins to work like the drawing of a make-believe car explains.

Of course, a real car has at least four cylinders, and more often six or eight. Any car that had only one cylinder would not run very smoothly. It would jerk ahead when the cylinder fired, and then pause until the piston came back up for the next cycle. When a car has several cylinders, each one fires at a slightly different time in the power cycle, and the engine and car run very smoothly. Have you ever been in a car that jiggled as the motor ran? This sometimes happens when the cylinders are firing out of order and their timing needs to be adjusted.

What makes this auto go?

2. The fuel pump, which is like a water pump, pumps the gasoline to the carburetor.

3. The carburetor, which is like an atomizer, changes the liquid gasoline into a gasoline and air vapor, which goes to the cylinder.

4. The cylinder is like a cannon with a piston in it.

5. The spark plug, which is like a lighter, ignites the vapor in the cylinder. The vapor burns and expands quickly, pushing the piston down.

6. The crankshaft is like the handle of a hand drill. The piston going down in the cylinder turns it.

7. The driveshaft is turned by the crankshaft, like a drill bit is turned by the drill handle.

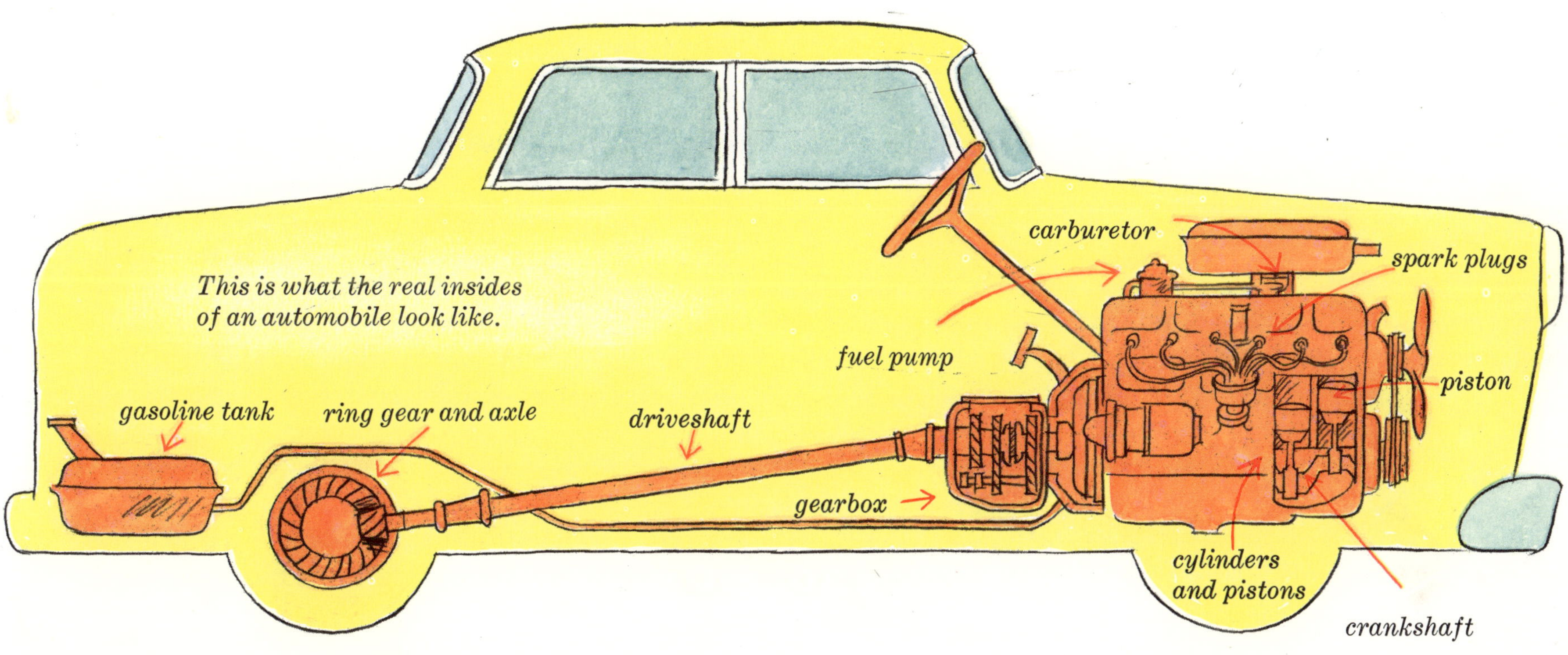

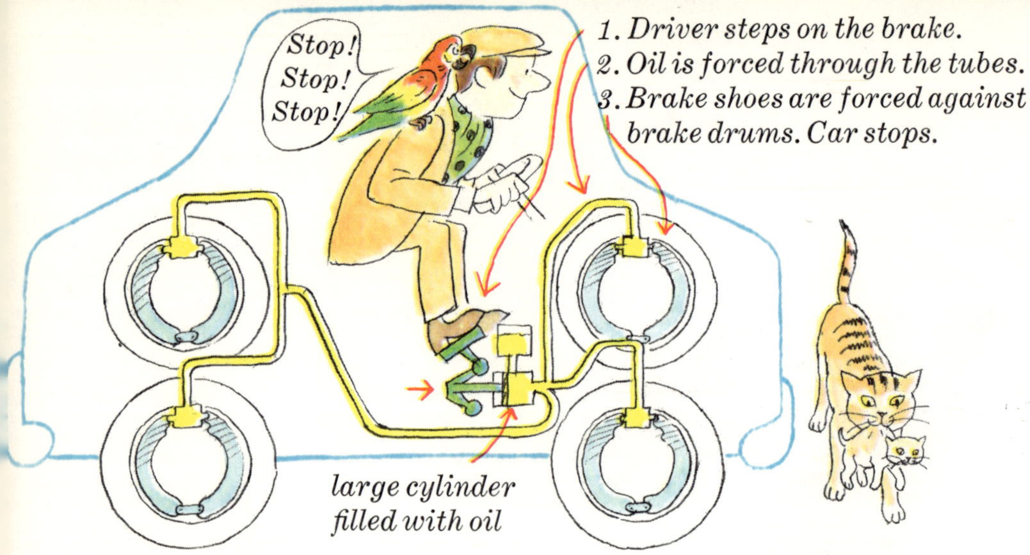

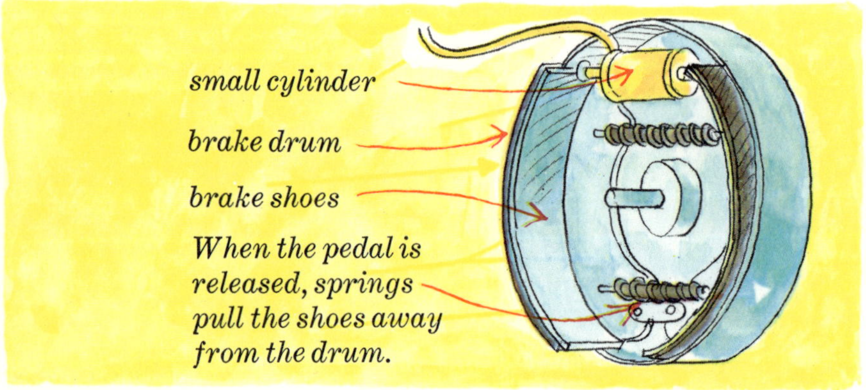

BRAKES. If you want to stop a soapbox racer, all you do is pull your stick brake against the edge of a wheel so they rub together, until the rubbing stops the wheel from turning. A real car needs brakes on all four wheels to stop it, and the driver needs a way to put on all the brakes at the same time. So there is a pedal for him to push, and all the stopping power is transferred to the four wheels at once, by a liquid! Here is how it happens.

The brake system is made up of cylinders and tubes, filled with oil. When the brake pedal is pushed down, it pushes a piston into a large cylinder of oil. Liquids cannot be compressed (squeezed) so the oil is forced out of the cylinder through the tubes to a smaller cylinder inside each of the four wheels. Each of these cylinders has two pistons, one at each end. The oil comes into the cylinder between the pistons, and pushes them apart. They push the brake shoes against the brake drum, which is turning with the wheel. The drum and the wheel slow down, until they finally stop.

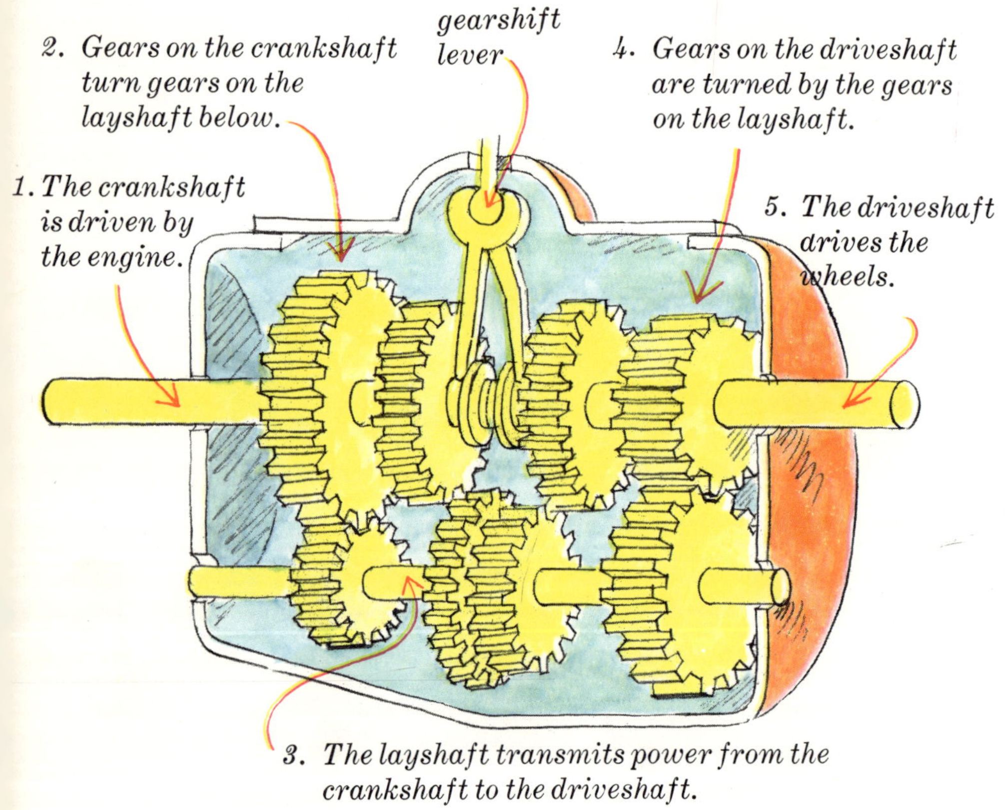

The TRANSMISSION is a set of different-sized gears which transfer power from crankshaft to driveshaft. (The crankshaft turns the gears, which turn the driveshaft and the rear wheels.)

Gears are needed because the engine is most powerful when it is running fast. This means the crankshaft is often turning faster than you want the driveshaft and wheels to turn. So, the gears are used to let the engine run fast, while they turn the driveshaft at slower speeds. What speed the driveshaft has depends on the sizes and arrangement of the gears (what gear the car is in). First is for starting off, second is for speeding up, and third is for cruising along. Reverse is for going backwards.

To change gears, the driver pushes the clutch pedal down, disconnecting the crankshaft from the gears. He uses the gearshift lever to shift the gears to a new position, and lets the clutch pedal up.

Reverse: going backwards

First: starting off

Second: picking up speed

Third: speeding along

WINDSHIELD WIPERS were invented so the driver could see the road ahead even through heavy rain or snow. Imagine a snowstorm before there were windshield wipers. The driver would stop the car, jump out into the storm, wipe the windshield, jump in again (soaking wet), drive for a few minutes, then do it all over again.

Windshield wipers are two arms that go back and forth across the windshield. Each has a soft rubber blade attached to it that wipes rain or snow off the windshield.

How do they work? A small motor under the dashboard, running on current from the battery, supplies the power. The motor turns a small wheel, which has a pin sticking out of its face near the edge. The pin goes into the ends of two extension arms. As the wheel goes around, the pin pushes these arms back and forth. The extension arms are attached to the windshield wiper arms which also move back and forth, and the rubber wipers clean the windshield.

The STEERING system controls only the front wheels of the car. No matter which way they turn, the rear wheels follow. The picture shows a simplified steering system. The steering wheel is mounted on the steering column. At the other end of the column is a gear.

The teeth of the gear fit into a row of teeth on a bar. The bar connects the pivot arm of one wheel to the pivot arm of the other. When the steering column gear turns, its teeth push the bar to one side. As the bar goes to the side it moves the pivot arms a little and turns the wheels. For instance, if the steering wheel is turned to the right, the steering column gear pushes the connecting bar to the left and the bar moves one end of each pivot arm to the left. This moves the other ends of the pivot arms (and the wheels) to the right. And the car turns to the right.

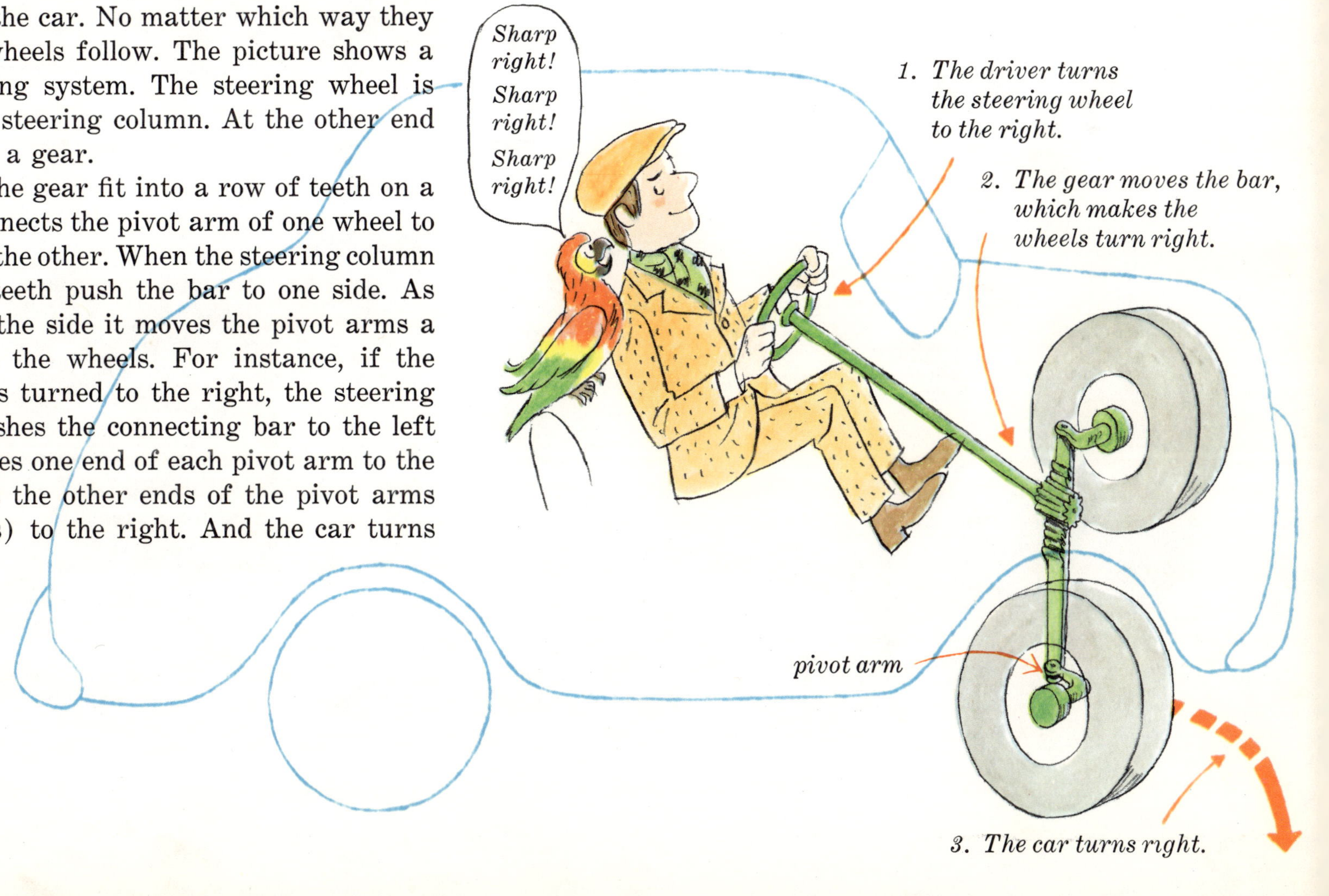

LOCOMOTIVES on rails were once run by horses and sails. But steam power worked much

As time went on, locomotives were improved and were able to pull longer trains and run faster.

As the locomotives grew longer and still stronger, they rushed faster and faster along the tracks.

better. Early locomotives look small and funny to us, but they did run and even pulled a few cars.

The boilers became bigger and more wheels were added. Trains traveled across whole countries.

"Old 999" went 112 miles an hour in 1893. The first little locomotives had zipped along at about 4.

A STEAM LOCOMOTIVE uses steam to push the pistons that drive its wheels. Because of a special valve called a slide valve, the steam pushes the pistons first backward, then forward. In the picture, steam goes into the cylinder through hole A, pushing the piston back. The slide valve is attached to the piston and slides back with it, blocking off A and opening hole B. Now steam goes in through B, pushing the piston and slide valve forward. B is blocked off, and the steam goes in through A again.

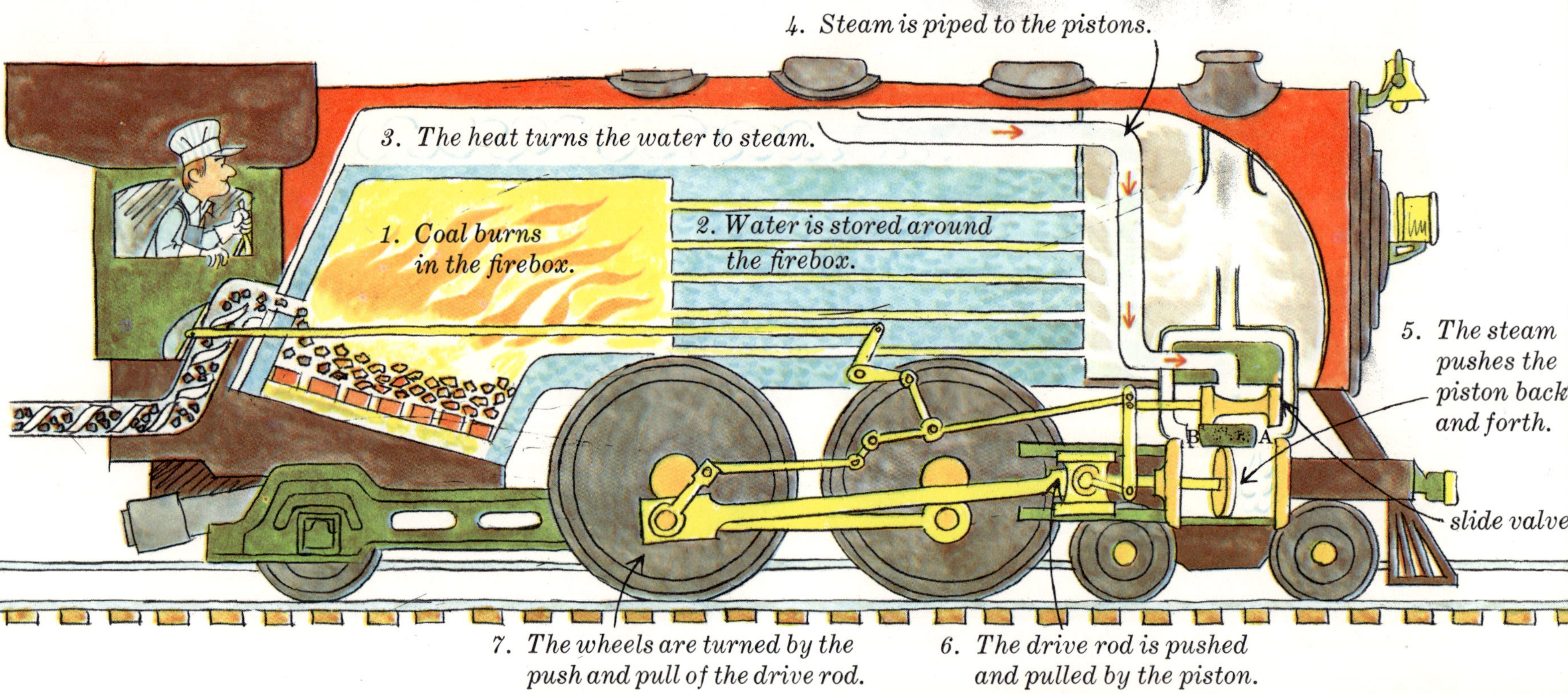

A DIESEL LOCOMOTIVE engine has cylinders and pistons, and is quite a lot like the engine that drives a car (see page 12). The main difference is that a diesel engine does not need spark plugs to start the fuel in the cylinders burning. As the pistons in the diesel engine move up in the cylinders, they squeeze the air until it gets very hot. (When gases like air are compressed, they get hot.) Then the diesel fuel sprays into the cylinders, and the hot air starts it burning explosively. The temperature and pressure in the cylinders rise, and the high pressure pushes the pistons down hard. As they move down, the pistons turn a crankshaft and a driveshaft.

In some diesels, the driveshaft powers the wheels directly. In the one in the picture, it runs a generator which sends electricity to small electric traction motors. They drive the locomotive wheels. This combination diesel-electric locomotive doesn't need overhead wires for power, yet it still can use electric motors, with their fast, smooth acceleration.

This ELECTRIC LOCOMOTIVE runs by electricity which comes to it through overhead wires. The current from the wires has too high a voltage for the traction motors, so a transformer in the cab brings it down to the right voltage. The current then goes to the traction motors, which power the drive wheels. These electric motors provide quick, smooth acceleration, so electric locomotives are ideal for pulling commuter trains, which must start and stop every few miles.

At first men paddled little BOATS made of hollowed-out logs, or pieces of bark or animal skins.

When a big sail was added, the blowing of the wind helped the rowers go still faster and farther.

Steam engines were added to the sails, first for riverboats, then for ships that could cross oceans.

Then, because they wanted to go faster and travel farther, many men would row a single boat.

Later, as bigger SHIPS were built with more and more sails, rowers were no longer needed.

Finally, ships didn't need sails at all. With steam power alone, they could reach any port on earth.

An OCEAN LINER is a sea-going city. The big ones carry over 2,000 passengers and 1,000 or more crew members. There are cabins for all the people, dining rooms, movie theaters, a laundry, barber and beauty shops, playrooms, even a hospital, everything to keep the passengers comfortable and happy. The ship carries all its own fuel and supplies, and makes its own heat and electricity. A metal ship with all this equipment weighs many tons. How can it float?

Think of a heavy metal pan. If you put it into the bathtub, part of it goes down into the water, displacing some of the water (pushing it aside). Long ago, a scientist discovered that an object in water is pushed up by a buoyant force equal to the weight of the water that it displaces. Since the pan

is hollow, it probably weighs less than the water it displaces. So, the buoyant force will hold it up, making it float. If the pan weren't hollow, however, it would weigh more than the water it displaces, and the buoyant force would not be enough to hold it up. A shipbuilder designs a ship to weigh less than the water it displaces, so it floats. As long as water can't get inside to weigh it down, the ship will keep floating.

Most modern liners have steam turbine engines. Steam from a boiler whistles through a turbine, which has blades like a fan. The steam pushing past the blades makes them turn, and they turn the propeller shafts and the propellers. The propellers bite into the water as they turn, pushing it out behind them and moving the ship forward.

Sailing in the same direction as the wind is called "running."

Sailing into the wind is called "sailing to windward."

Sailing across the wind is called "reaching."

A SAILBOAT uses wind for its power. If you are satisfied to let the wind blow you where it chooses, sailing is easy. But a good sailor makes the boat go where he wants it to go. If the wind is blowing from behind, there is no problem. The wind fills the sails, and pushes them and the boat forward.

If the wind is coming from the side, it's a little different. The centerboard keeps the boat from slipping sideways. The sails catch some of the wind, and since the boat cannot go sideways, it is pushed forward, but not as fast as when the wind is coming from behind. If the wind blows right at you it is impossible to go directly forward. You must zig-zag forward so that the wind pushes first from one side of the sail, then the other. This is called "tacking."

You steer the boat with the tiller, which moves the rudder to the side. The water flowing past the boat pushes on the rudder, and the boat turns.

A SUBMARINE is a completely watertight metal ship that can float on the surface of the water or dive below and ride along like a fish. When it is on the surface it uses a diesel engine to turn its propeller. When it is under water, it switches to battery-driven electric motors, since the diesel engines need air to run. Some subs use nuclear reactors for power. The reactors don't need air so they can be used under water, too.

A submarine has large tanks built into it, called ballast tanks, filled with air. To make the submarine go under water, the men open valves on the outside of the tanks, allowing the air to escape. Then water comes in to replace the air. When the tanks are full, the submarine weighs exactly as much as the water it displaces. The buoyant force no longer holds it up, so it slips below the surface. Now it can use its special fins to steer down and up under water. To surface, the men steer the sub up near the surface. Then the water is blown out of the ballast tanks with compressed air and the sub becomes lighter than the water it displaces. Up it goes.

Men wanted to fly and thought of bird-like machines and air-carriages pulled by birds in harness.

They built GLIDERS and sailed them in the sky. Only the air currents kept them from falling.

Planes were improved and flights grew longer. One man, all alone, flew across the Atlantic Ocean.

They made BALLOONS filled with hot air and flew in football-shaped DIRIGIBLES.

With engines and propellers added, real AIRPLANE flying began. But the trips were short.

Then came HELICOPTERS and planes with many engines that carried people everywhere.

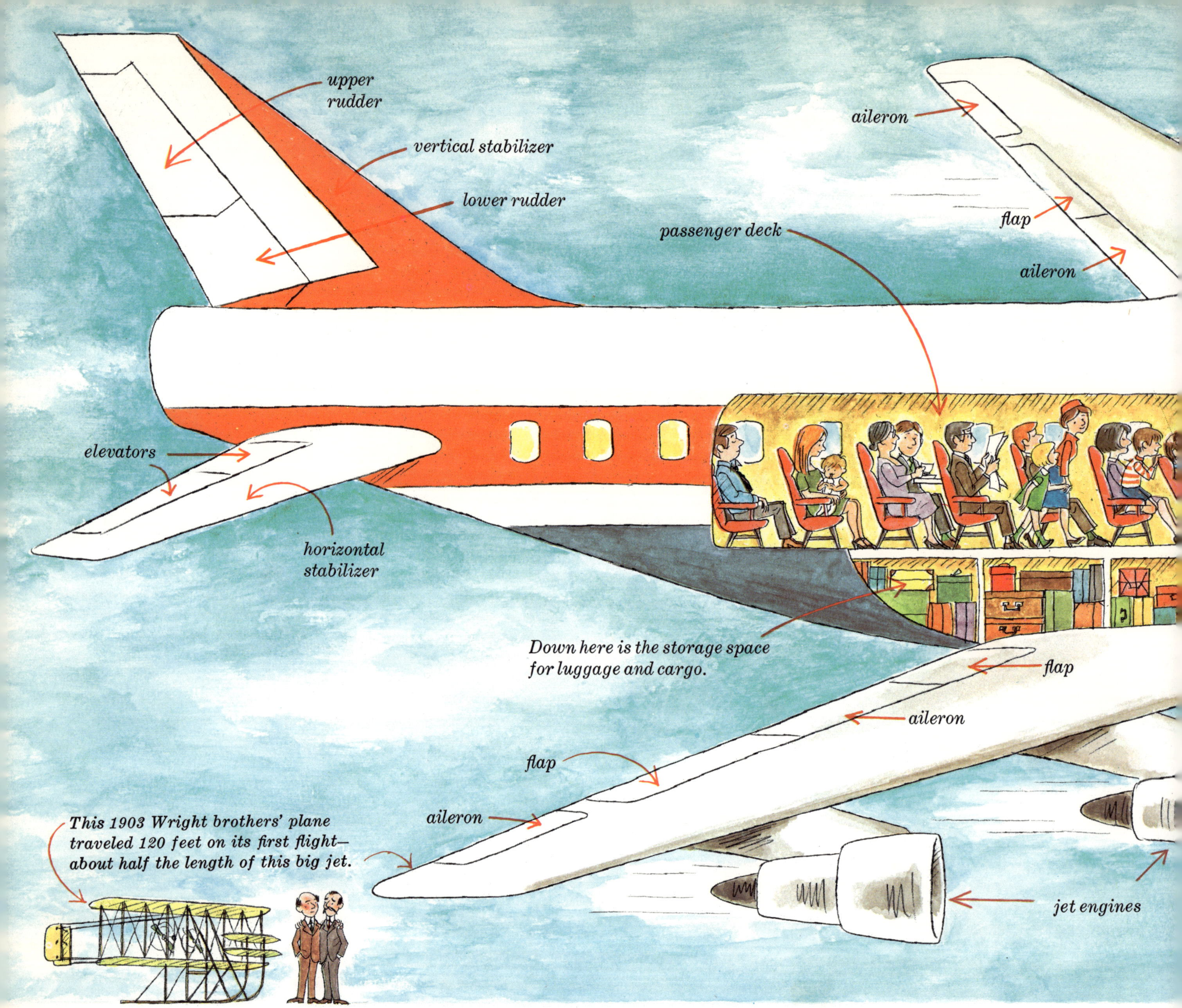

Two main forces make a plane fly. THRUST makes it go forward. It comes from the plane's engines.

LIFT keeps the plane up. The wings' shape causes the air under the wings to have more pressure than the air on top. The higher pressure keeps the plane from falling.

The JET AIRPLANE is the fastest means of transportation in everyday use. This large jet carries about 350 passengers, 15 crew members, and all their luggage. It also has a kitchen, one or two movie screens, a passenger lounge, fuel tanks, and cargo space. Four turbofan engines attached to the wings provide the power. The diagram shows what happens in the engine.

The engine is pushed forward through the air the same way a blown-up balloon is when you let

go of it. In the balloon, air shoots out the back, pushing it forward. In the turbofan engine, the hot mixture of burning gas and air shoots out the opening in the back, pushing the plane forward.

What keeps a heavy plane like this one up in the air? Air pressure does most of the work, and the shape of the wings makes it possible. The upper side of the wings is curved, and the underside is nearly flat. This is so that the air rushing over the top of the wing will travel farther and faster than the air rushing under it. Fast moving air has a lower pressure than slower moving air, so the air pressure over the wings is lower than the air pressure under them. The higher pressure under the wings pushes the plane up and keeps it flying. This is why the plane has to race down the runway before it can take off. It has to get the air moving past the wings, so that the right air pressures are set up.

The next page explains how the ailerons, elevators, and rudder of a plane work.

Remember? Blow up a balloon—let it go—and

—just like a jet!

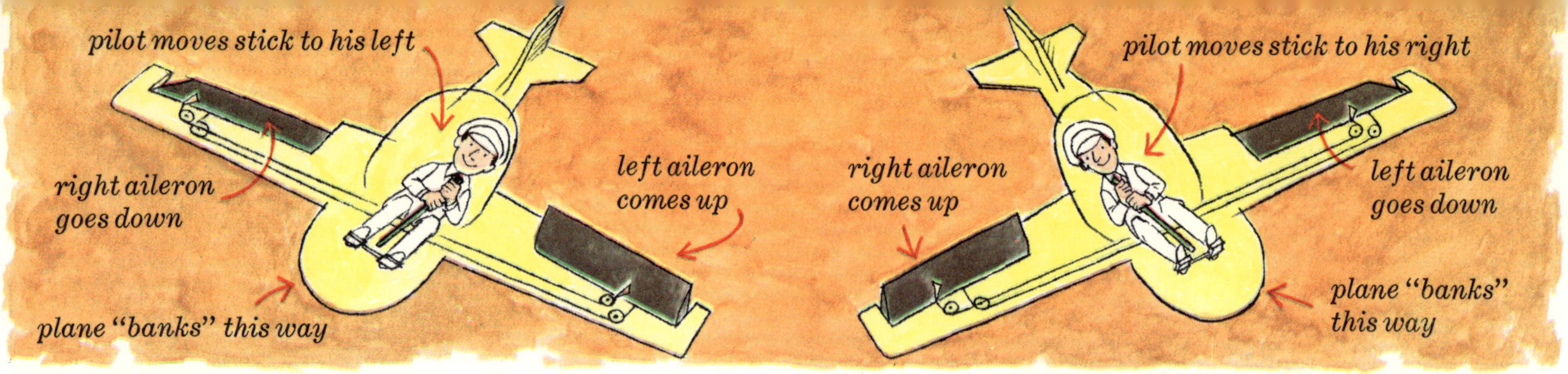

The AILERONS are used to "bank," or tip the plane sideways. To bank to his left, the pilot moves the stick to the left. The left aileron comes up. The air going over the wing hits the aileron and pushes against it, tipping the left wing down. Meanwhile, the right aileron goes down and the air going under the wing pushes against it, tipping the right wing up. The plane banks to the left.

To bank to his right, the pilot moves the stick to the right, and everything works in just the opposite way. Usually, the pilot banks because he wants to turn, and to do that, he uses the rudder, too.

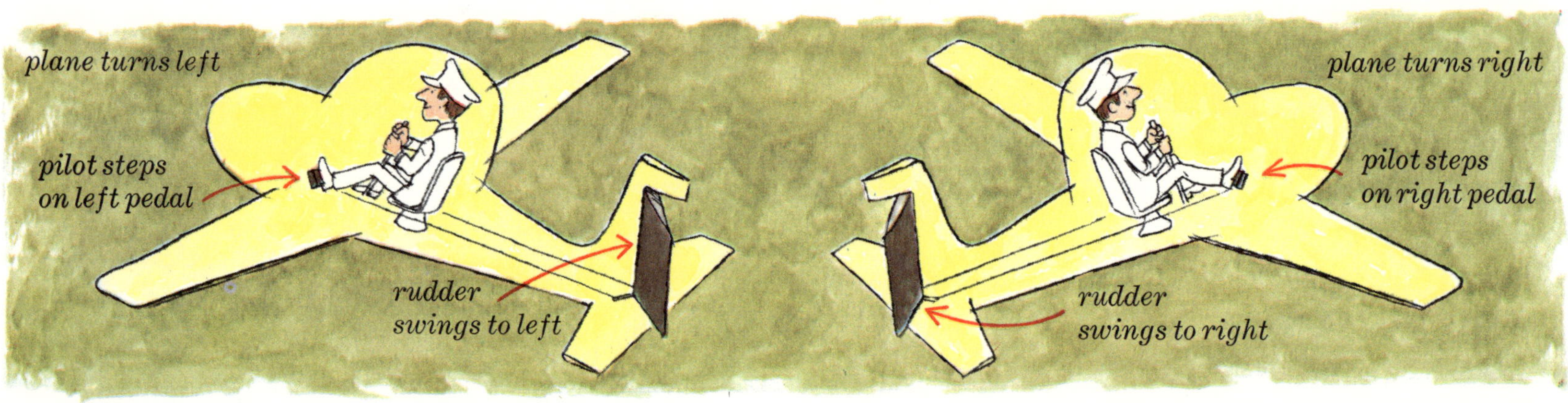

The RUDDER is used with the ailerons to turn the plane. To turn left, the pilot banks to the left. At the same time, he steps on the left pedal, which moves the rudder to the left, out into the stream of air rushing past the left side of the tailfin. The air tries to push the rudder back, moving the whole tail of the plane to the right. So the nose of the plane goes to the left and the plane turns left. To turn right, the pilot banks to the right, and moves the rudder to the right. You can probably see that the rudder, like the ailerons, uses the air flowing past the plane to steer it.

The ELEVATORS also use the stream of air to change the plane's position. They make the plane go down (dive) or up (climb). When the pilot moves the stick forward, the elevators go down. The air flowing under the plane's tail pushes against them, forcing the tail up, and the nose down. The plane dives. To climb, the pilot pulls the stick back, moving the elevators up. The air flowing over the plane's tail pushes against them, forcing the tail down, and the nose up. The plane climbs.

It is with these simple controls, the ailerons, rudder, and elevators, that a pilot flies every kind of plane, from a tiny private plane to an enormous cargo transport.

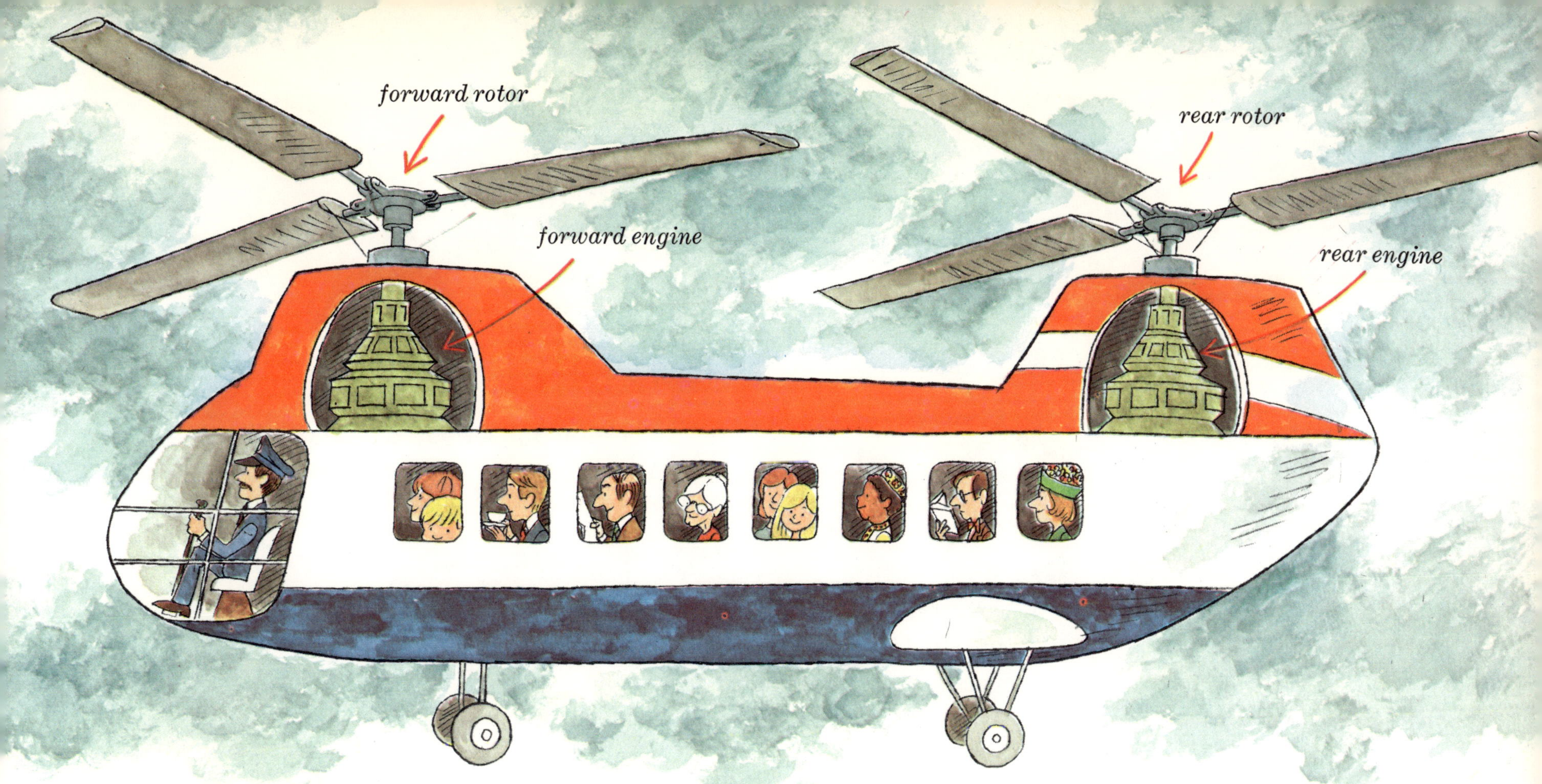

A HELICOPTER can fly in any direction, up, down, sideways, forward, or backward. The helicopter's rotors do the same job as the wings, ailerons, propeller, and rudder of the airplane.

The rotor blades are really the helicopter's wings. They are shaped like wings. When they whirl around, air is forced over and under them. The air moves over the curved top of the rotor blade very fast, creating a low pressure area above the blade. The higher pressure under the rotor pushes it and the helicopter up. That is why helicopters don't need to race down a runway to take off. They don't have to move forward to get the air rushing past their wings. All they have to do is spin the rotor blades.

Some helicopters have little tail rotors to keep their bodies from starting to spin under the big rotor once they are in the air. Bigger helicopters have two large rotors. The front rotor spins in one direction, and the rear rotor spins in the other. This also keeps the helicopter stable in the air.

The rotor blades can be tilted forward and backward. By tilting them certain ways with his hand and foot controls, the pilot can make the helicopter turn and go up, down, forward, and backward.

A helicopter can rise straight up and come straight down, so it needs just a small landing space.

It can fly along very slowly and even hold still in the air—handy for astronaut pick-ups.

Because helicopters fly this way, policemen can use them to watch out for trouble below—like a traffic jam.

The SATURN ROCKET has five stages. The first three are full of fuel tanks and rocket engines. They put the service and command modules into orbit around the earth.

When a rocket engine is started, fuel from one tank is combined with liquid oxygen from another and they go into the combustion chamber where a small charge of gunpowder explodes and starts them burning. (The oxygen is there so the fuel can burn.) Hot gases from the fire rush down and out the rocket nozzle, pushing the rocket up.

Rocket engines are very powerful and they need a lot of fuel. That is why a rocket is built in stages. After the fuel in the first stage is gone, that stage separates from the rocket. This gets rid of a lot of weight—the rocket engines, the empty fuel tanks, and the first stage itself. This means the second stage has a much lighter rocket to push up into the sky. When the second stage has used all its fuel, it, too, separates from the spacecraft, and the third stage takes over. About twelve minutes after lift-off, the third stage engines shut down and the spacecraft goes into a parking orbit around the earth.

Now the astronauts check over all their equipment, to make sure everything is working correctly. If it is, they start up the engines of the third stage again, to push them out of earth orbit and send them on their way to the moon. Finally, they have to get the lunar module (LM—*lem*) out of the third stage. They open the doors over the LM, turn the command module around with their thruster rockets, and dock with the LM. Slowly, they pull the LM out, and move away from the third stage.

Guess which one of these astronauts was the first man to step on the moon?

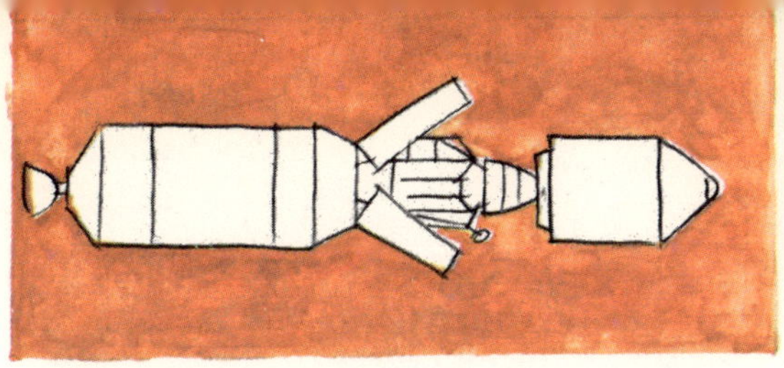

The doors that protect the LM open.

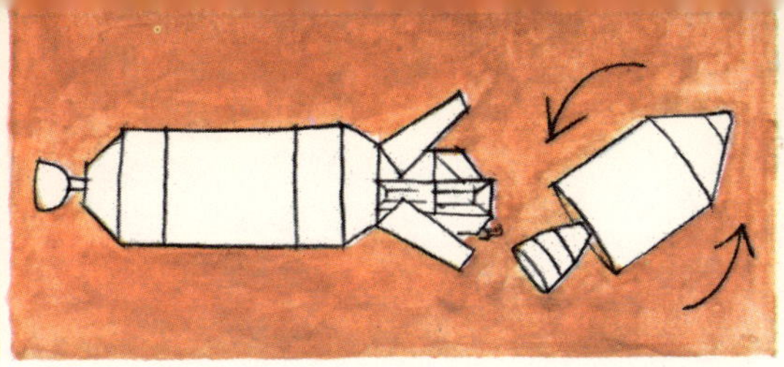

The command module turns around.

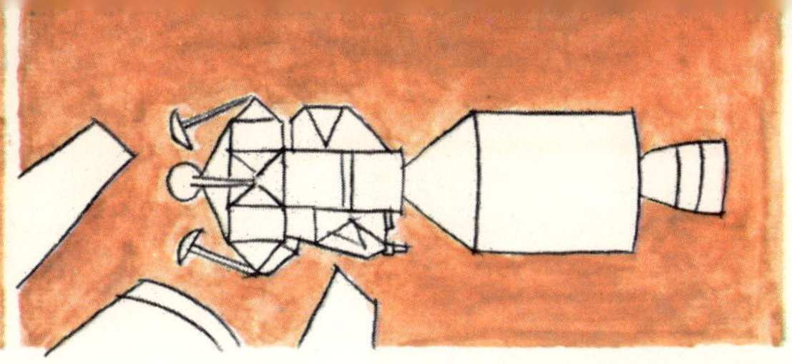

The command and lunar modules dock.

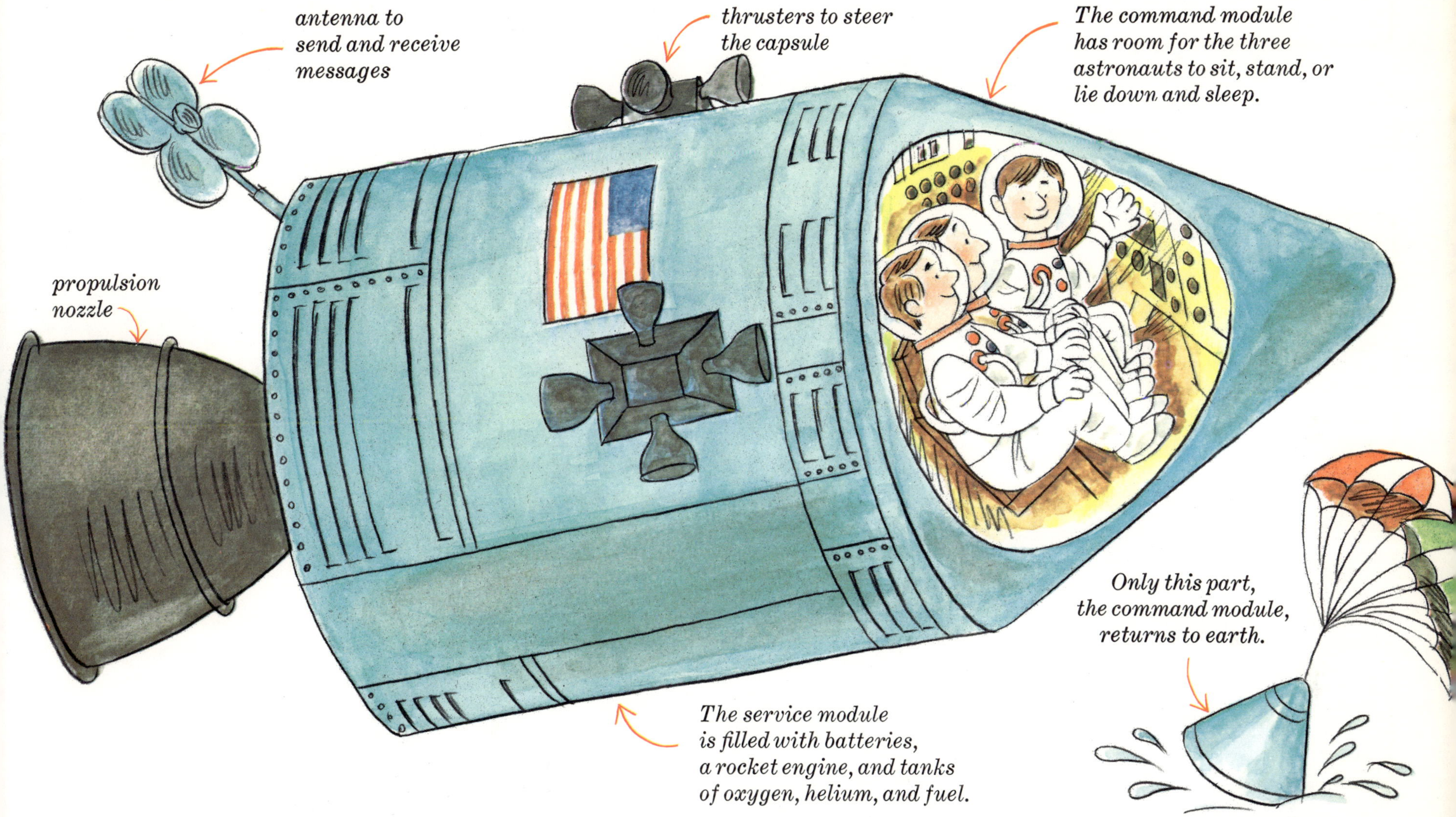

The APOLLO CAPSULE is made up of the service and command modules. The service module supplies electricity and oxygen to the command module, and makes the crew's drinking water.

The command module is the crew's quarters, and carries their food, and a small supply of oxygen and water for re-entry. Its life-support system keeps the temperature at a comfortable level, cleans and freshens the circulating oxygen, and keeps the capsule pressurized, so the astronauts don't have to wear their spacesuits all the time.

The astronauts are in constant contact with the base by radio and television. Tiny medical instruments attached to their bodies radio back their heartbeats, blood pressures, and temperatures.

Getting from the earth to the moon and back again is complicated, because everything in space is moving. The moon is revolving around the earth, while the earth is revolving around the sun. The astronauts have to take off from the one moving object and aim for a point in space where their path will cross the path of the other moving object. Long before the voyage, computers figure out where that point will be, and what course and speed will get the capsule there at the right time. During the voyage, computers at the base and aboard the command module constantly figure and refigure the positions, courses, and speeds of the capsule, moon, and earth, and indicate if mid-course corrections are needed. They also watch over the complicated systems that run the spacecraft. No man could do all this. Without computers, spaceflight would be impossible.

The LUNAR MODULE (LM) carries food, water, and oxygen for the astronauts, and its own fuel, a computer, a heating and cooling system, and the tools the astronauts will use on the moon.

After the command module goes into orbit around the moon, two of the astronauts switch to the LM. When they have checked their instruments, they undock from the command ship.

Using the descent-stage rocket engine, the pilot moves the LM out of orbit. The moon's gravity pulls on the LM, and it begins to fall in toward the moon. Now the pilot uses the descent engine to slow the LM's fall, and guide it to a gentle touchdown on the moon. The astronauts begin their experiments.

When their work is all done, the astronauts are ready to lift off. The ascent engine pushes them back into orbit. Now the LM and the command module use radar signals, special homing signals, and their computers to find each other. They make small course corrections with the thrusters, and complete their docking. Soon they will be ready to fire the service module's engine to start the long trip home.

The SPACESUIT keeps the astronauts alive in space, where unprotected men would die. It has several layers. First, under the suit, the astronaut wears a one-piece water-cooled suit with a network of tiny tubes through which cool water circulates, picking up excess heat from his body.

The spacesuit must be pressurized, so it has a rubberized airtight pressure layer. Oxygen comes in under this layer from the PLSS (*pliss*) back-pack and pressurizes the suit. The suit's outer layer protects the astronaut against the extreme heat and cold of space, and keeps any tiny meteorites from piercing the pressure layer.

The gloves attach at the wrists and the helmet clamps on at the neck. Under the helmet, the astronaut wears a cap with built-in headset and microphones, so he can communicate with the base and his partner. The PLSS back-pack holds the receiving and transmitting equipment.

A MAGNET attracts iron, and is usually made of steel or iron. One end of a magnet is called its north pole and the other is called its south pole. The north pole of one magnet attracts the south pole of another, but two north poles push each other away. So, poles that are different attract each other, and poles that are the same push each other away.

Magnetism is caused by a special arrangement of the atoms in the magnet. Iron atoms are like little magnets. In a plain iron bar they are all in a jumble, and their magnetism cancels out. But in an iron bar magnet, the atoms are arranged with their north poles all pointing one way and their south poles pointing the other. Now their magnetism adds up and makes the bar one big magnet.

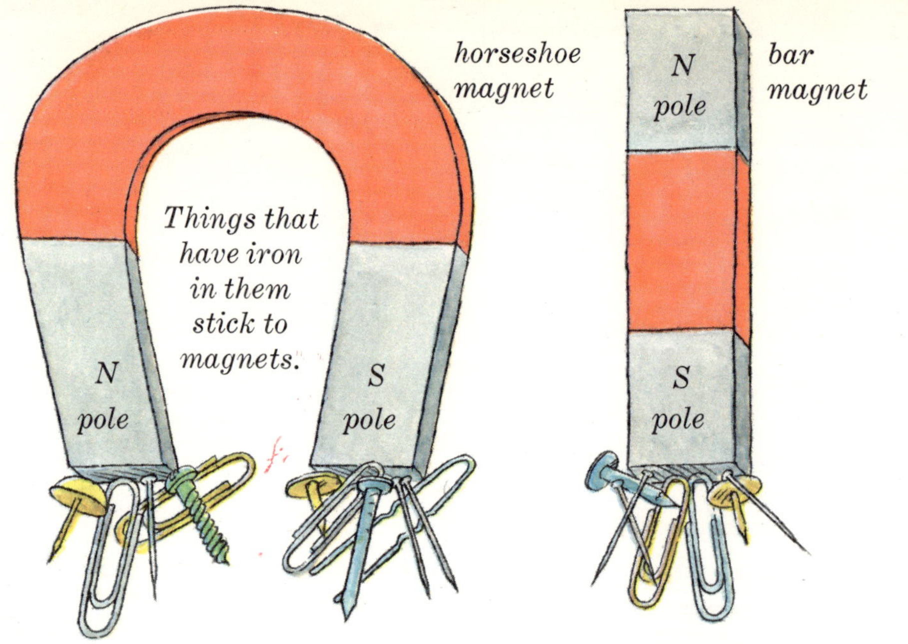

Things that have iron in them stick to magnets.

A magnet will not pick up most things that don't have iron in them.

Let a bar magnet hang down on a string. One end always points north and is called the north pole of the magnet. The other end, of course, points south. It is called the south pole.

A compass needle is a magnet, so it points north and south.

Compasses help boy scouts find their way in the woods...

...captains steer their ships across the oceans...

...and pilots guide their planes through the clouds.

An ELECTROMAGNET is a piece of iron magnetized by an electric current. This can happen because electricity running through a wire creates magnetic forces around the wire. For instance, if you wrap a wire around an iron nail and run electricity from a battery through the wire, the magnetic forces will arrange the atoms of iron in the nail so that their north poles all point the same way. The nail becomes a magnet, until the current is turned off. While the nail is magnetized, it has a north and south pole just like any magnet. Which end is north depends on which direction the current is flowing through the wire. If the direction of flow is reversed, the poles reverse, too.

Current is running from the battery through the wire. The nail is now an electromagnet.

This large electromagnet is strong enough to pick up heavy scrap iron.

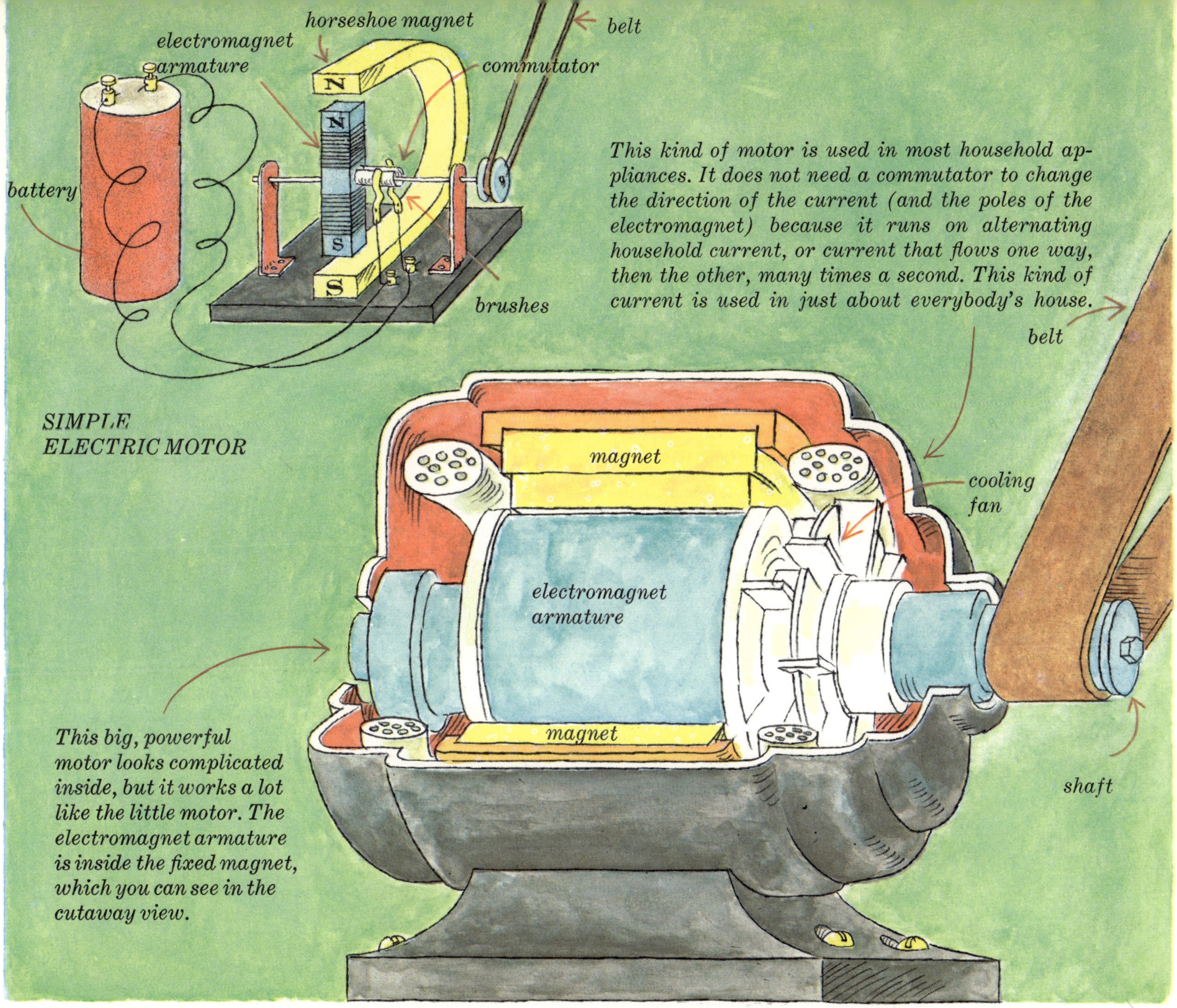

This kind of motor is used in most household appliances. It does not need a commutator to change the direction of the current (and the poles of the electromagnet) because it runs on alternating household current, or current that flows one way, then the other, many times a second. This kind of current is used in just about everybody's house.

This big, powerful motor looks complicated inside, but it works a lot like the little motor. The electromagnet armature is inside the fixed magnet, which you can see in the cutaway view.

Most ELECTRIC MOTORS work a lot like the simple one at the top of the picture. In this simple motor, an electromagnet is mounted on a shaft between the poles of a horseshoe magnet. When the motor is turned on, electric current runs through the wire, the brush, and the commutator into the electromagnet. In the picture, the north pole of the electromagnet is next to the north pole of the horseshoe magnet. The south poles are next to each other, too. Because poles that are the same repel each other, the armature (the electromagnet and its shaft) will turn.

You can probably see that after the electromagnet has made a half turn, it will tend to stop because the north and south poles of the two magnets will attract each other. To keep this from happening, the commutator acts like a switch, reversing the direction of the current flowing around the electromagnet at just the right time. This reverses the magnetic poles, and, just as when the motor was first started, poles that are the same are together. They repel each other, and the armature keeps turning. The poles change with each half turn the armature makes, and so the motor keeps running. A belt attached to it can be used to run different machines.

How many things can you think of that are run by electric motors?

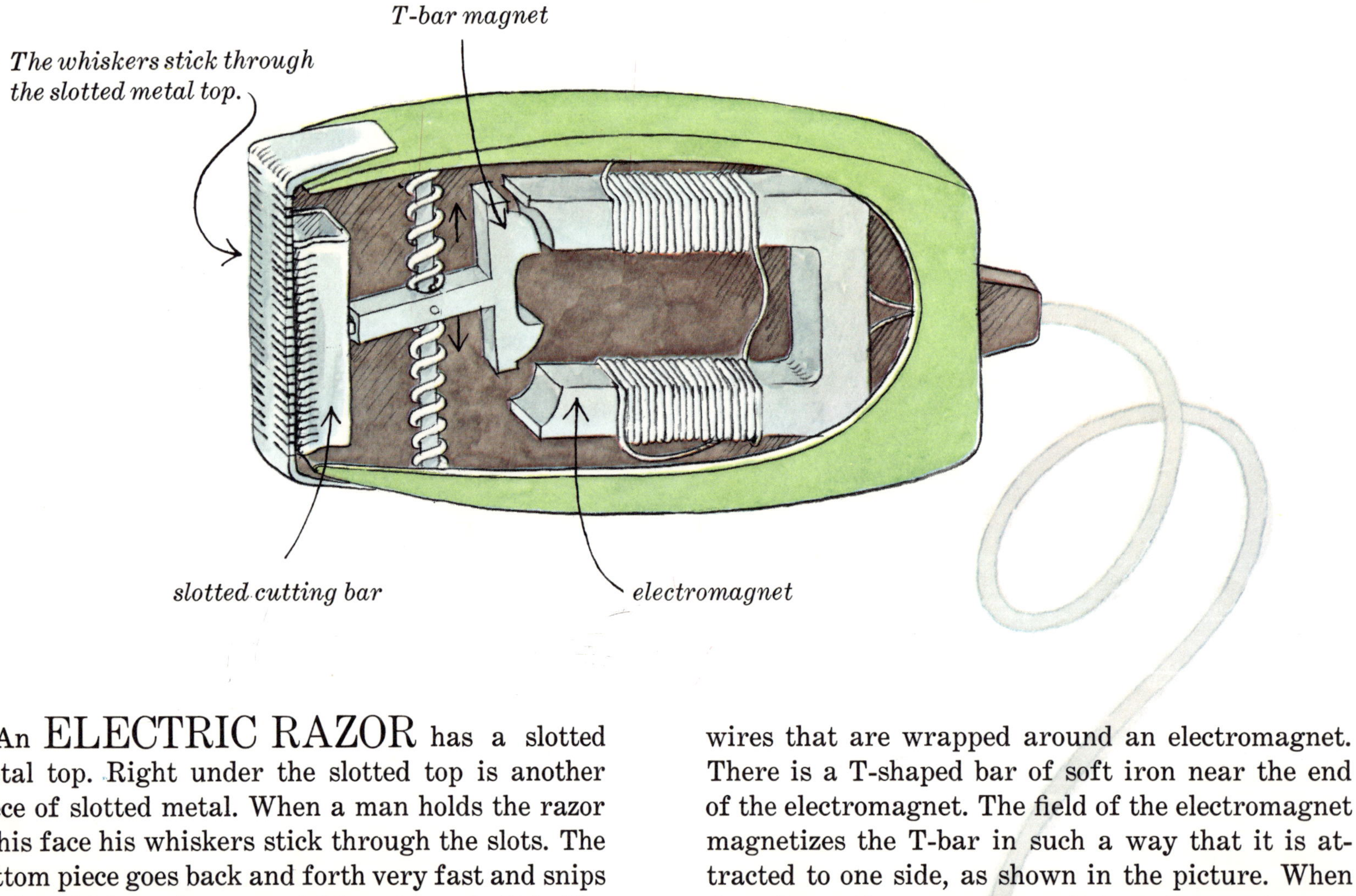

An ELECTRIC RAZOR has a slotted metal top. Right under the slotted top is another piece of slotted metal. When a man holds the razor to his face his whiskers stick through the slots. The bottom piece goes back and forth very fast and snips off the whiskers that are sticking through. It is like the action of many, many tiny scissors.

An electric motor makes the cutter go back and forth. Electricity comes into the razor and goes to wires that are wrapped around an electromagnet. There is a T-shaped bar of soft iron near the end of the electromagnet. The field of the electromagnet magnetizes the T-bar in such a way that it is attracted to one side, as shown in the picture. When the alternating current switches the north and south poles of the electromagnet, the T-bar flies over to the other side. This happens many times a second, and moves the slotted cutter back and forth.

The ELECTRIC TOOTHBRUSH

swivels back and forth very fast. Inside the fat handle of the one in the picture is the small motor that makes it go. The motor gets its power from a battery also in the handle. Other electric toothbrushes get their power from a regular electric socket.

A system of gears and levers changes the spinning motion of the motor's shaft into the back and forth motion needed for the toothbrush. A little gear on the shaft turns a larger gear which has a little post sticking up from it. As the post goes around in a circle, it moves a lever back and forth. The lever swivels the toothbrush back and forth, and your teeth get brushed the right way . . . automatically!

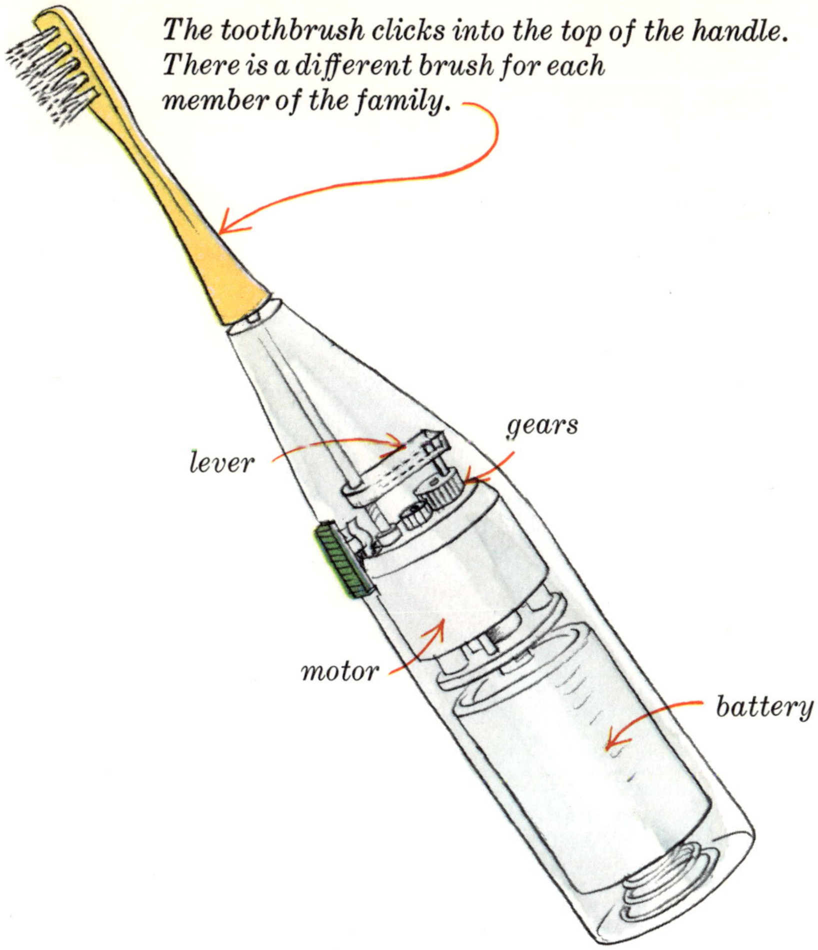

A jetstream **TOOTH CLEANER** cleans the spaces between your teeth with little jets of water. It has a small electric motor to run a little water pump, which pumps the water out of the tip in spurts. To use the tooth cleaner, you turn the lid over, fill it with water, and place it on top of the machine. The hole in the bottom of the lid sits on a tube that leads to the pump. From the pump a thin rubber hose leads to the tip, which you hold in your hand.

When you turn the machine on, the motor moves a lever back and forth. This lever moves the piston back and forth in the pump's cylinder. When the piston goes back, it draws just a little water down from the lid into the cylinder. When it pushes forward, it pumps that bit of water through the hose and out the tip. It pumps so fast that hundreds of spurts of water come out of the tip each minute.

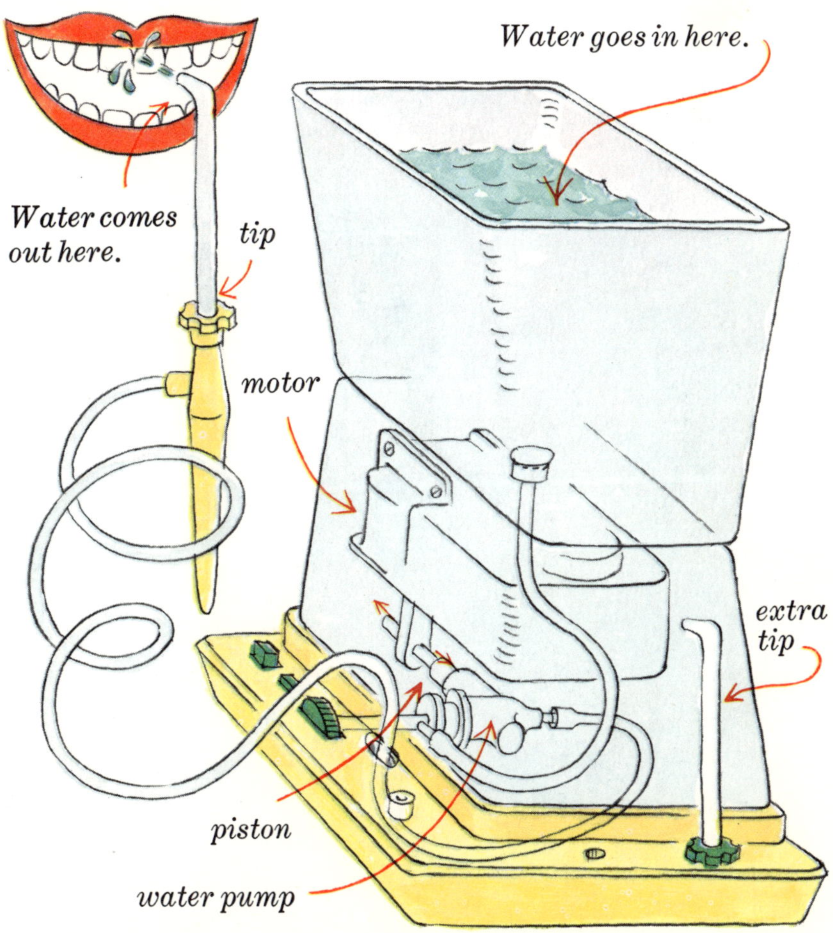

The first step in making a dress is cutting out the pieces from the cloth.

Here are the pieces ready to be sewn on the machine.

And here is the finished dress.

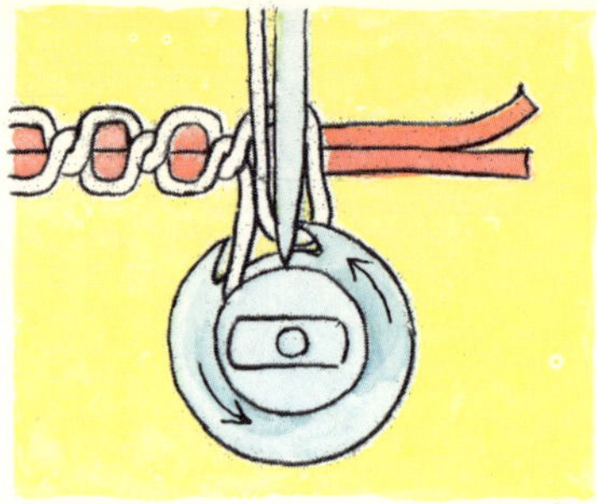

1. *The needle goes down through the cloth and the hook on the turning bobbin catches the needle thread.*

2. *The bobbin keeps turning and the hook pulls the needle thread, looping it around the thread from the bobbin.*

3. *The needle goes up, and the take up lever goes up, pulling the needle thread tight and forming a stitch.*

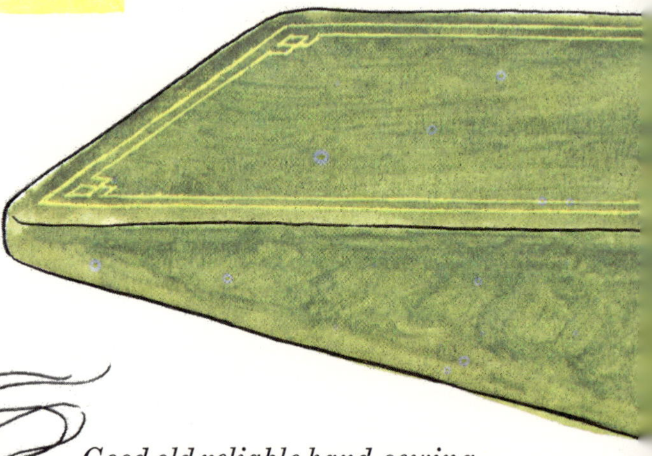

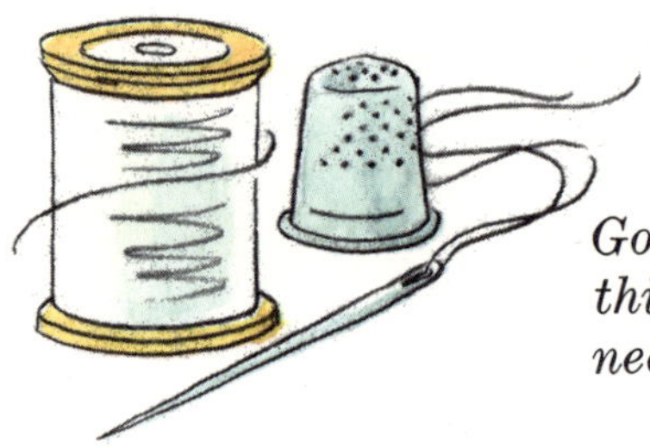

Good old reliable hand-sewing things are still used for jobs that need special care or stitching.

A SEWING MACHINE makes hundreds of stitches in a minute. It uses two spools of thread. One spool sits on a spool-pin on top of the machine. The thread winds around and passes through several thread guides and finally threads through the needle. The other spool of thread, called the bobbin, is inside the machine beneath the needle. The bobbin thread goes up through a hole in the sewing platform.

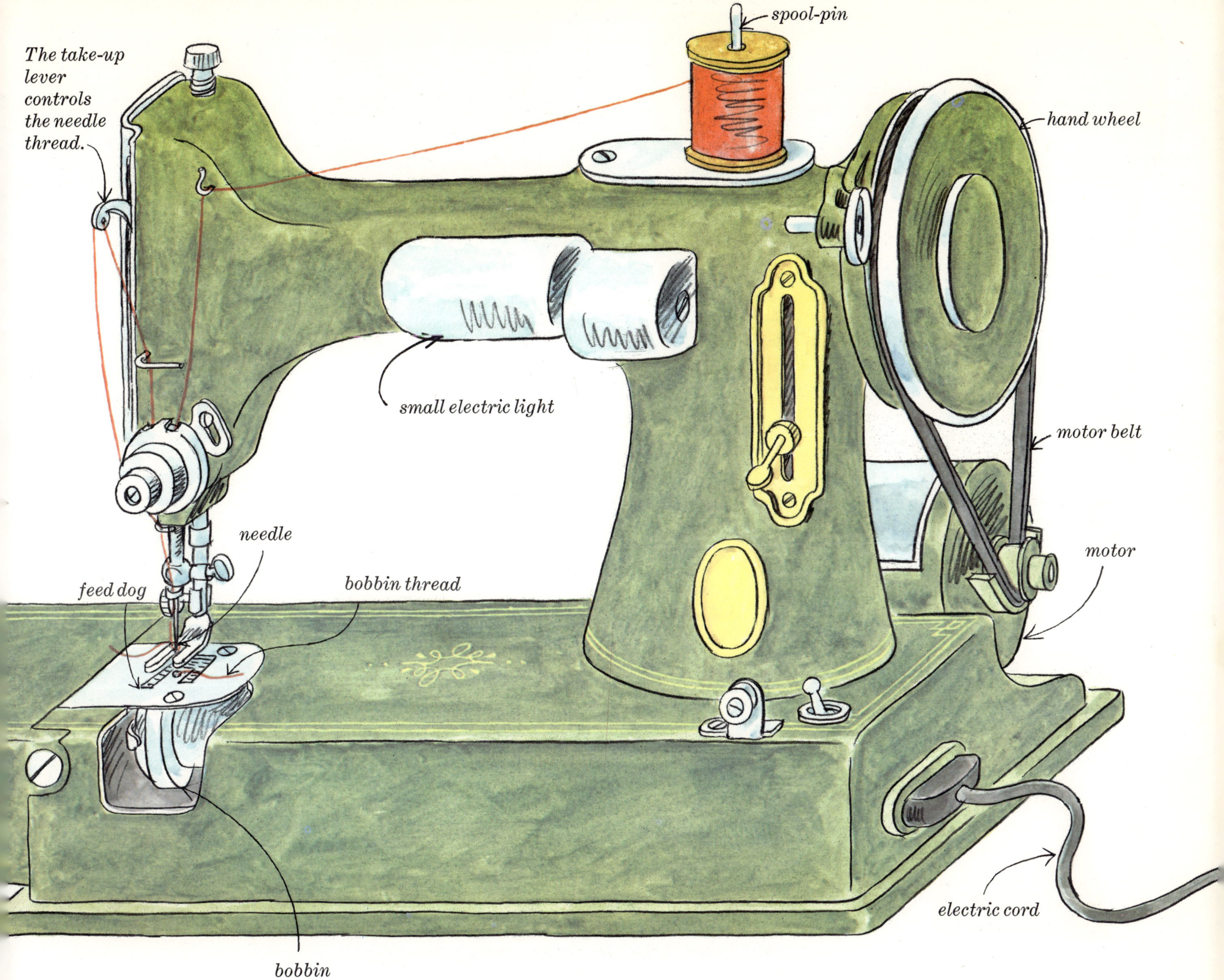

This is what happens when you press the foot pedal. Electricity makes the motor go, and the motor belt turns wheels and gears inside the machine that move the needle up and down, and turn the bobbin around and around.

Now comes the hard part—how the two threads join together to make stitches. First, the threaded needle goes down through the cloth. When the needle is down below, a hook sticking out of the bobbin catches the needle thread and loops it around the bobbin. When the needle goes back up, the needle thread forms a tight loop around the bobbin thread. That is one stitch.

After each stitch a gadget with little saw-teeth comes up under the cloth and pushes it ahead a bit to make room for the next stitch. This gadget has a funny name . . . it's called a feed dog because it feeds the material through the machine.

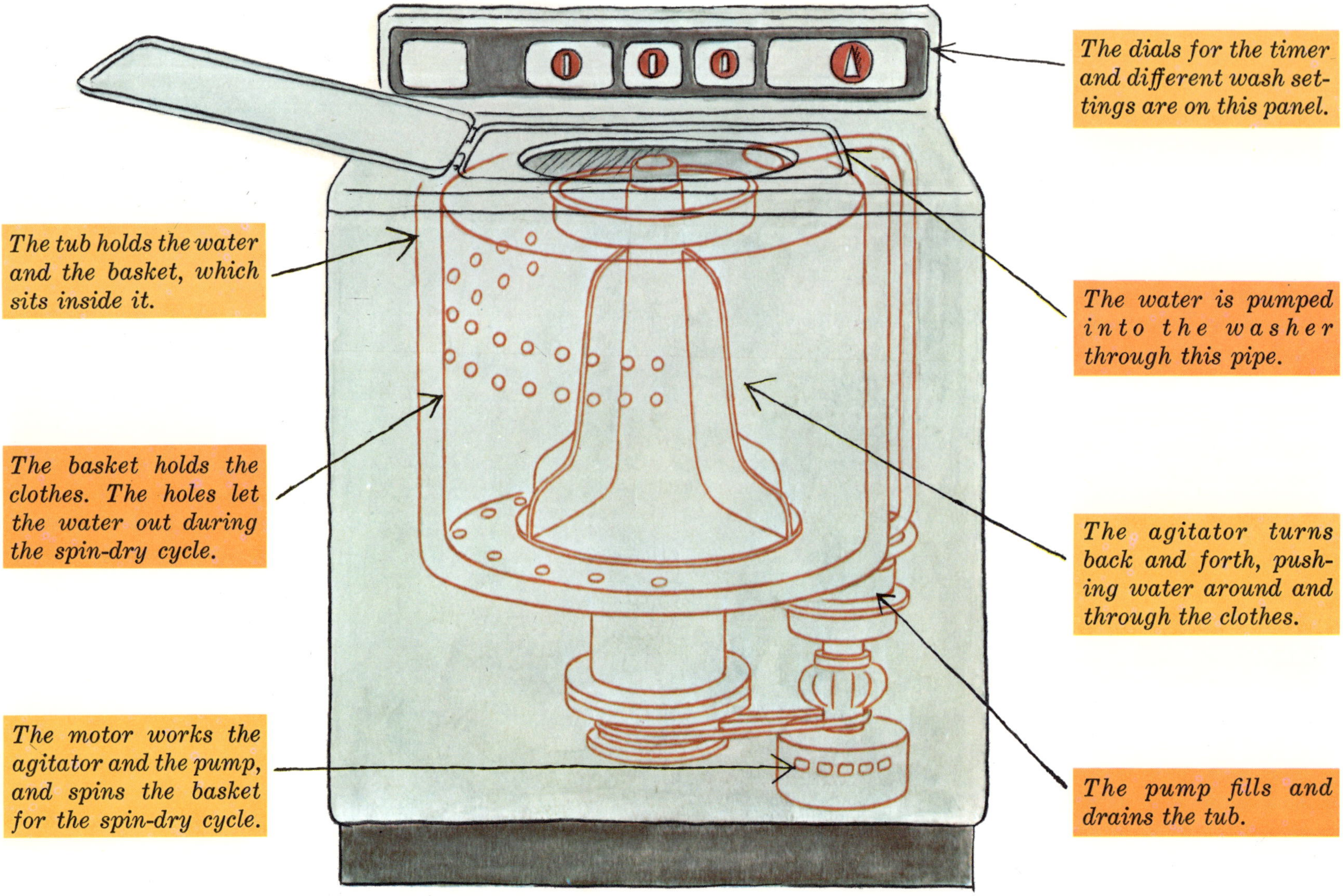

A WASHING MACHINE fills itself with water, washes and rinses the clothes, drains out the water, and turns itself off, all automatically. The timer in the washer controls all these things. All you have to do is put in the clothes and the soap, turn on the machine, then take out the clothes when they are clean.

Hot and cold water pipes lead into the machine. When you turn on the washer, valves are opened and water is pumped into the tub until the timer shuts it off. As it ticks off the time, the timer operates switches which start and stop the soak, wash, rinse, and spin-dry cycles.

An electric motor in the bottom of the machine turns the agitator during the wash and rinse cycles and also spins the tub during the spin-dry cycle. The agitator churns from side to side or, in some machines, up and down. This churning action tosses the clothes around inside the tub and forces the soapy water through them until they are clean.

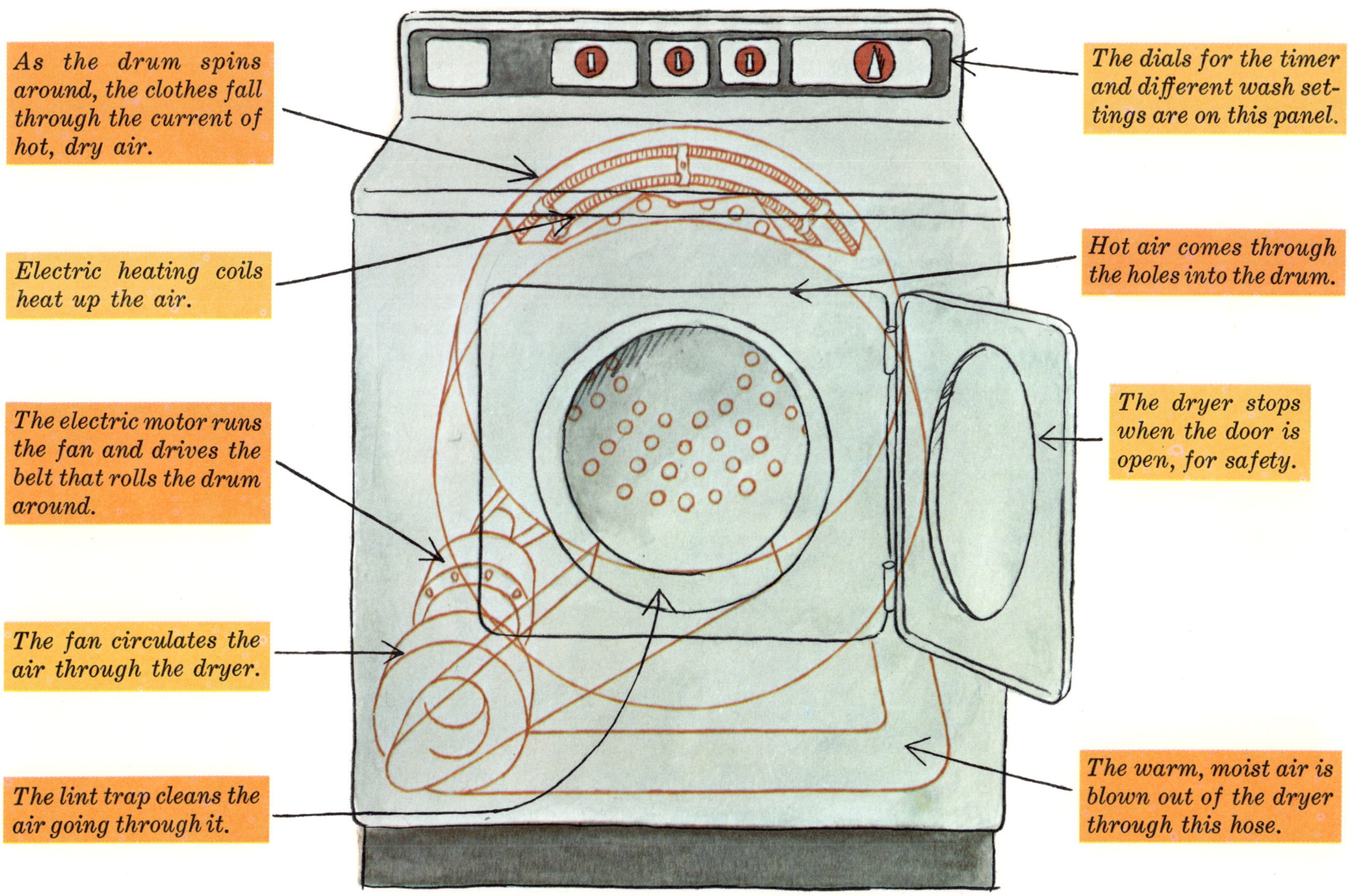

A CLOTHES DRYER heats air and uses it to take the moisture out of wet clothes and carry it outdoors.

A fan blows the hot air on the wet clothes while they tumble around in a rotating drum. An electric motor drives the belt that turns the drum. The drum keeps turning and the fan keeps blowing until a timer turns off the dryer.

The fan pulls air into the dryer, and draws it past the heating elements that heat it up, into the drum. The hot, dry air absorbs the moisture from the clothes and becomes warm, wet air. This wet air is blown out of the dryer through a special vent, carrying the moisture to the outdoors.

Some dryers have several temperature settings for different kinds of clothes. A thermostat keeps the dryer at the temperature you set and turns it off automatically if the air inside gets too hot for safety. You can find out how a thermostat works if you read about the electric iron, on page 52.

The old-fashioned broom is still in use.

*The rug beater . . .
hang up the rug
and whack it!*

*The feather duster
can reach high places.*

*The little whisk broom,
used to brush furniture
and clothes, sends dust
flying about the room.*

*The Shakers invented
the flat broom which
sweeps much better than
the old-fashioned one.*

In the tank-type vacuum cleaner, the fan is behind the dust bag instead of in front.

This broom-type vacuum cleaner is for floors and rugs.

This little vacuum cleaner hangs from the shoulder.

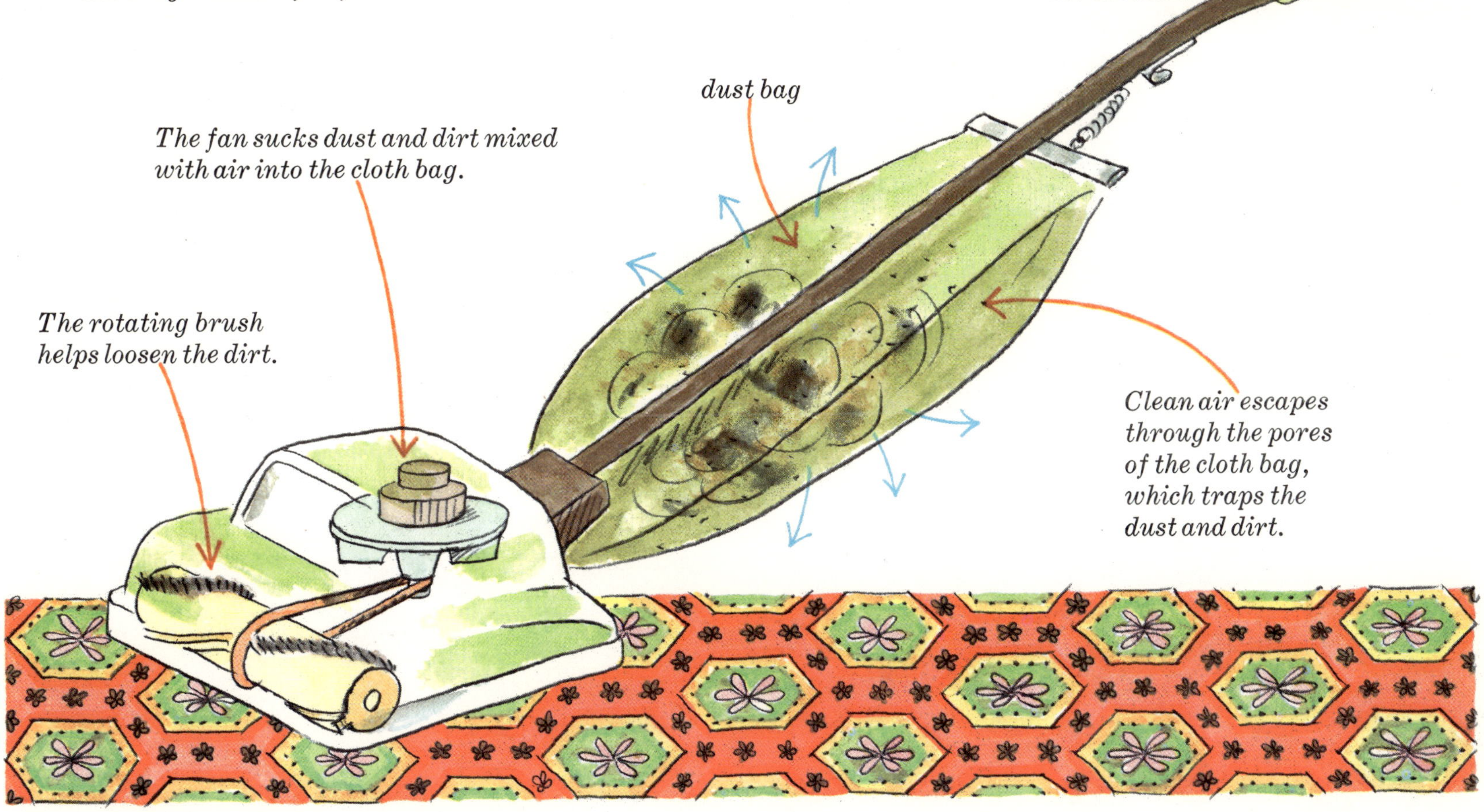

The VACUUM CLEANER is the modern broom. It sweeps floors and carpets, and catches the dust in a bag, instead of sending it into the air. There are many different sizes and types of vacuum cleaners, but they all work in the same kind of way. An electric fan sucks air and dirt through the front of the cleaner and into the bag. The air escapes from the bag through its pores. The pores are too small for the dirt to go through, so it collects in the bag.

The vacuum cleaner works because it has a partial vacuum inside. Anywhere there is less air than the usual amount, there is said to be a partial vacuum. You can make a partial vacuum in a pop bottle by sucking some of the air out with your mouth. Then you'll feel a pull on your lips because of the partial vacuum. This pull, or suction, is what the vacuum cleaner uses to pull things into itself.

When you turn on the vacuum cleaner, the fan creates a partial vacuum by pushing air out behind it. More air rushes in through the front to try to fill up the vacuum. As long as the fan goes, it keeps the partial vacuum going, and air keeps rushing in through the front, carrying dust and dirt with it.

An electric motor runs the fan, and in some vacuum cleaners, like the upright in the picture, it also turns a brush which loosens the dirt way down in the fibers of rugs.

An ELEVATOR is a small room that carries people up and down in buildings, from floor to floor. It is raised or lowered by steel cables attached to its top. The cables hang over the pulley shaft of an electric motor. The elevator hangs from one end of the cables and a heavy weight called a counterweight hangs from the other end. The elevator and counterweight ride up and down between steel guide rails, which keep them from wobbling about.

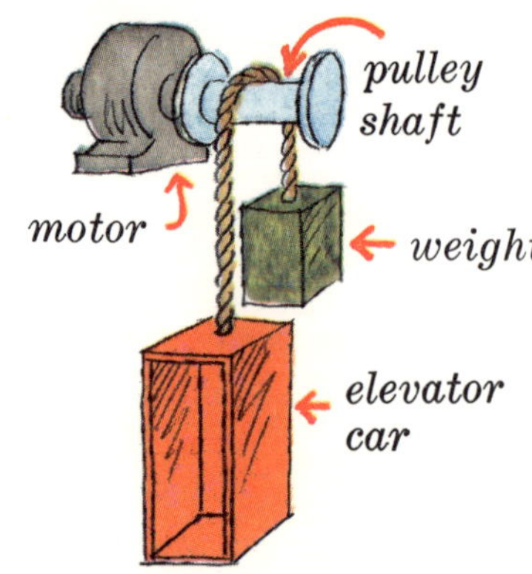

Most elevators today are so automatic that they don't need an operator to run them. It's all done with buttons that send electrical signals through wires to the motor. When you step into the elevator, you press the button with the number of the floor you are going to. After a mechanism closes the doors, the motor turns the shaft and the elevator starts. The motor cannot start until the doors are safely shut. When the elevator gets to your floor, it stops, and the doors open to let you out.

There are also buttons beside the elevator door on each floor, marked UP and DOWN. When you push one of them the motor at the top starts working and brings the elevator to your floor.

This is how a counterweight works. If you want to lift a heavy piano, and you have an elephant that weighs almost as much as the piano, you can rig it up this way. Then it will be easy to lift the piano, because the weight of the elephant pulls down on the rope and helps. You only have to do enough work to lift the difference between the weight of the elephant and the weight of the piano. Counterweights in elevators pull down on the cables and help the motor lift the car.

An ESCALATOR is a series of steps pulled up or down an incline by a chain. Each step has its own wheels that run along slanted rails. It's a kind of train being pulled up a hill with a passenger standing on each little car.

An electric motor supplies power for the escalator. The motor turns big gears with teeth. The teeth fit into the links of a chain, and they pull the chain and the continuous loop of steps up and around, then down and around.

At the top of an UP escalator, the steps flatten out and slide into a slot in the floor. This makes it easy to step off. The steps go all the way back down underneath the escalator, and come out a slot in the bottom, ready to go up again. A DOWN escalator works the same way, except that the steps move from top to bottom.

The motor also turns the wheel which moves the flexible handrail around at the same speed as the steps. The handrail helps you get on and off the moving escalator.

An ELECTRIC FAN is used to make people cooler, but it doesn't cool the air. It just keeps it moving by blowing it around. And when the air is moving around it helps moisture evaporate faster and that makes things cooler. For instance, on a hot day, after playing outdoors, you come home. There is perspiration on your face and the air from the electric fan hits it. The perspiration evaporates, taking heat from your face, and you feel cooler.

A fan works very simply. The motor and the fan blades rest on a metal stand. The fan blades are attached to the shaft of the motor. When you turn on the motor, the blades spin. They have a protective wire guard over them because the blades could hurt you if you put your hand into them when they're spinning. The blades are at an angle and when they turn they scoop up the air and push it forward . . . making a nice cooling breeze!

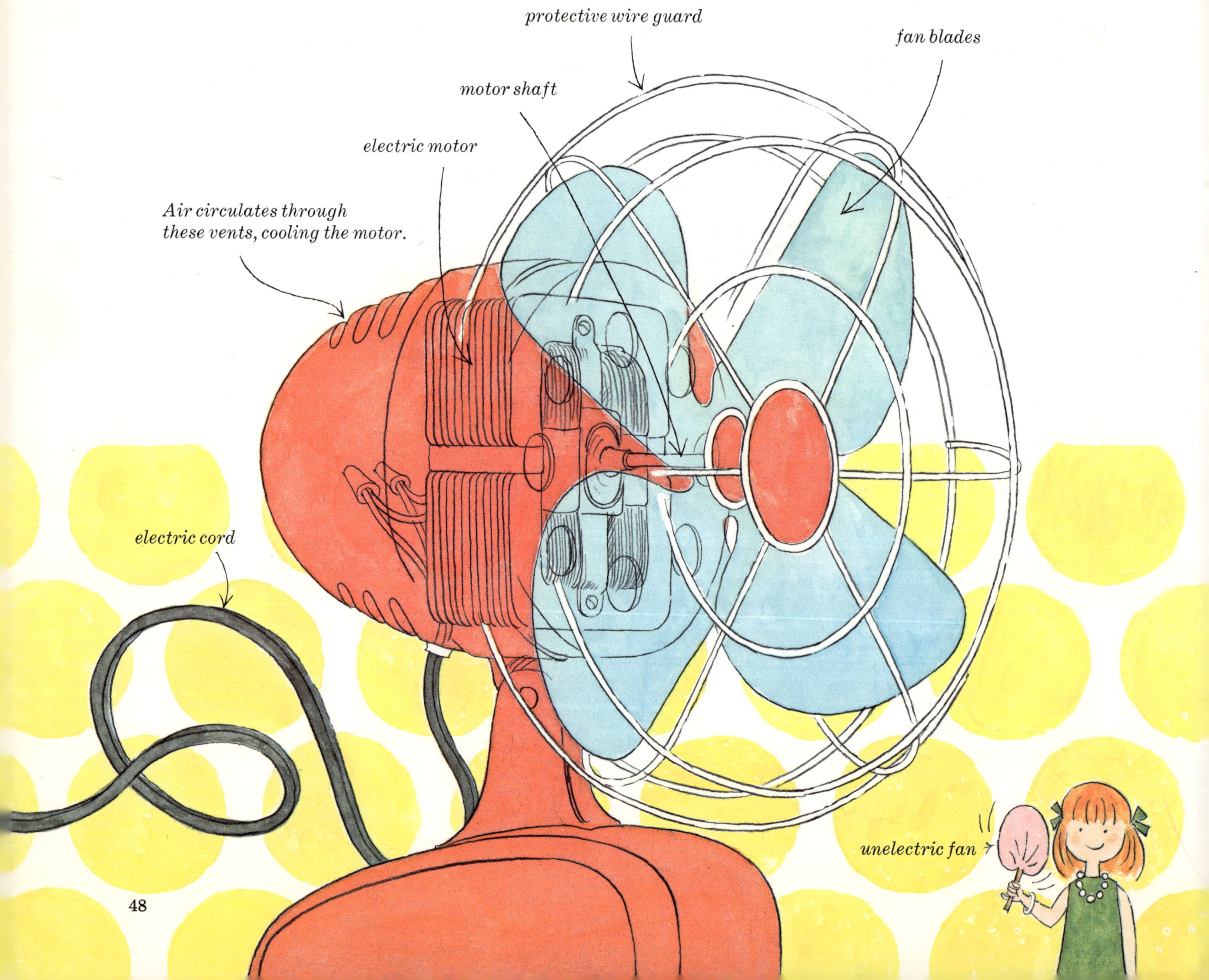

An AIR CONDITIONER takes heat from one place, say the living room, and puts it into another, usually outside the house. To do this, it uses the principle of evaporation. (Remember how the evaporation of perspiration cools your face?) But in the air conditioner things are a little more complicated. A liquid called a refrigerant runs through pipes and acts like a messenger, carrying heat from inside the house to outside.

In the hot room, the air conditioner fan blows the hot room air around the pipes. The refrigerant in the pipes evaporates (changes into a gas), absorbing the heat from the air, which gets cool and is blown back into the room. This is the cooling part of the cycle.

Now the hot refrigerant goes to the outside pipes where a compressor squeezes it, making it hotter, hotter even than the outside air. The cooler outside air takes the heat out of the refrigerant. The result of this is that the heat goes into the outside air, and the refrigerant turns back into a liquid. It is ready to start cooling your room again.

Not so long ago ice was cut from frozen lakes.

Then the iceman carried it through the streets and sold it. "Ice! Ice!" he called.

The iceman put the cake of ice right into the icebox.

Today a REFRIGERATOR doesn't use ice to keep itself cold. Instead, it uses a motor and a system of pipes that contain liquid refrigerant. That is why it is called a refrigerator. Here is what happens to the refrigerant during the refrigeration cycle. It starts out as a liquid, traveling up the pipes on one side of the refrigerator. When it reaches the coils of the freezing unit the liquid evaporates. As it evaporates, it absorbs the heat from the air in the freezing unit and the refrigerator.

Now the hot refrigerant gas travels down the other side. At the bottom, a compressor squeezes it and a fan blows cool air past the pipes. The gas loses the heat it picked up in the freezer compartment and changes back to a liquid. As you can see, the refrigerator is really just another air conditioner, cooling a very small room which happens to be an insulated box for keeping food.

This diagram shows the refrigerant system.

Do you think an Eskimo needs a refrigerator?

A SKATING RINK is cooled very much like a refrigerator. The refrigerant is carried through pipes buried inside the concrete base of the rink. Water is flooded on top of the concrete, and a motor begins pumping the refrigerant liquid through the pipes. It evaporates and takes the heat from the concrete and the water. The water turns into ice. This is the cooling part of the cycle.

Now the refrigerant must get rid of the heat it picked up. It runs through pipes behind the rink. The cooler outside air flows around the pipes, taking the heat from the refrigerant. It turns back into a liquid, ready for the next cooling cycle. And, while all this is going on under the ice, you can be skating around on top.

The ELECTRIC IRON uses a heating element just like the stove does. The element is inside the heavy iron base or sole plate. An electric current goes to the heating element through a thermostat, which controls the temperature of the iron.

The thermostat uses a bimetal strip, made up of two strips, each of a different metal, joined together. Metals expand when heated, and when this bimetal strip is heated it curls, because the two different metals expand different amounts. The one that expands the most bends over the other one. When the temperature of the iron goes over the level you've set it for, the bimetal strip is hot enough to bend away from its electrical contact. The flow of electricity stops, and the iron begins to cool. As it cools, the bimetal strip straightens until it touches the contact again and the electricity starts to flow. You set the temperature by turning a dial which moves the contact closer to or farther away from the bimetal strip.

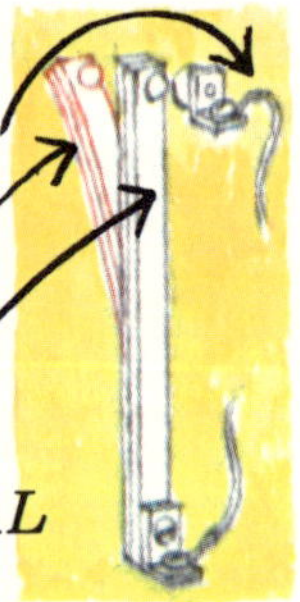

BIMETAL STRIP

The iron in the picture is a steam iron. It has a small tank of water inside. The water drips into an open section of the sole plate, where the heat of the sole plate changes it to steam. The steam shoots out the holes in the bottom of the iron, and softens the wrinkles in the clothes, so the hot iron can press them out.

Here are some other things that work a lot like the iron and toaster. All of them have a heating element that gets hot when an electric current passes through it.

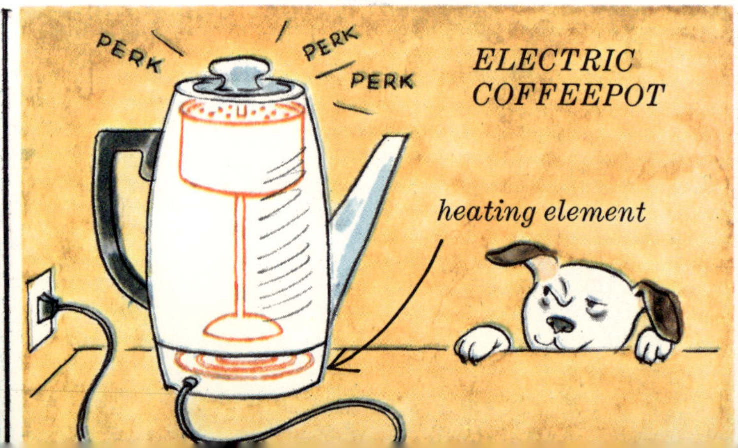

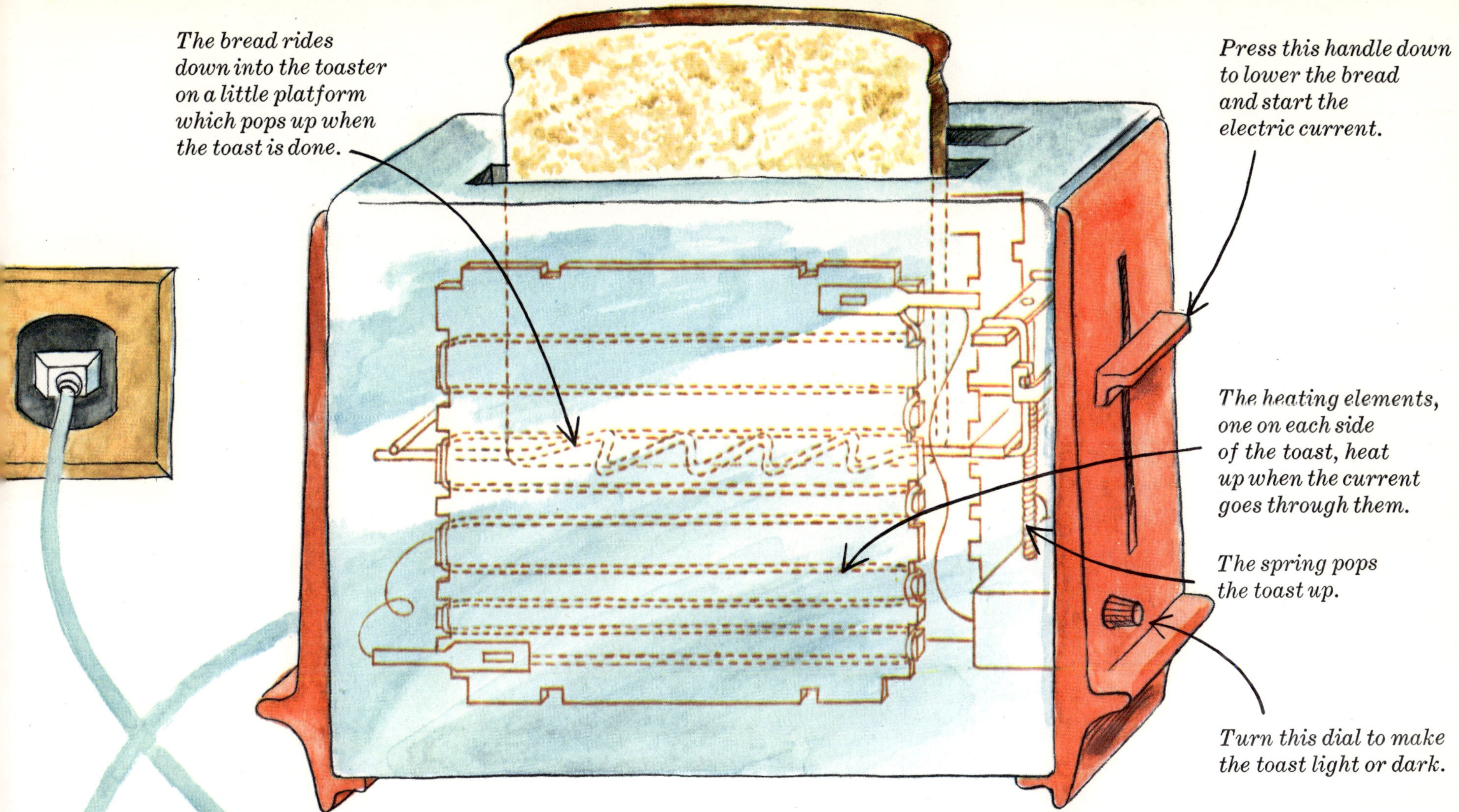

The TOASTER uses very thin heating elements, strung on thin sheets of a heat resistant material on either side of each toasting slot.

You probably think that making your breakfast toast is very simple, and it is, for you. You just put the bread into the toaster, set the dial for how brown you want the toast to be, and press the handle down. The bread goes down. Then the toaster takes over, and a lot happens down there.

The handle you pushed down catches on a special switch and connects the house current to the heating elements. The elements glow red hot as the current runs through them, and toast the bread. The best toasters have a thermostat and a timer to control how brown the toast gets. In these toasters, the current runs through a bimetallic strip on its way to the heating elements. When the heating elements have warmed up to toasting temperature, the bimetallic strip has warmed up enough to curl over and push a button that starts the timer. The timer, which you have set with the light-dark dial, ticks away a certain amount of time, and then releases the handle when that time is up. A spring pops the handle and the toast up.

The heating elements go off because the handle no longer connects them to the source of the electric current. And you have a slice of toast, done just the way you like it.

A **GAS STOVE** cooks and bakes with the flames of burning gas. The gas comes into the stove from pipes in the wall. Metal tubes take it to each stove-top burner ring. Gas has to mix with air before it can burn, so there is an airhole in each tube. Under the center of the stove top is a pilot light, a small gas flame which burns all the time. When you turn the burner knob on, gas starts flowing to the burner. Some of this gas goes down the tube to the pilot light, which ignites it. The flame rushes back up the tube and lights the burner. There is also a large burner ring under the oven. When it is lit, hot air rises into the oven through holes in the oven floor.

Some gas ovens automatically control their temperature by means of a thermostat. (See page 52.) When the oven gets too hot, the bimetal strip bends to cover a valve, shutting off the gas flowing to the oven. As the oven cools off, the strip straightens to open the valve again.

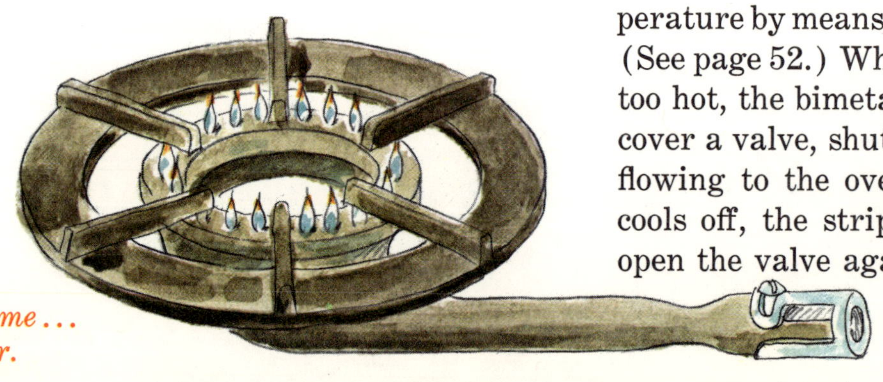

A gas burner heats with flame . . .
a big flame cooks faster.

An ELECTRIC STOVE has heating elements for its burners and oven. When you turn the burner knob on, electricity runs through the wire in the core of the heating element and heats it up very hot. The element is surrounded by insulation, which keeps the burner from getting too hot. A metal tube covers the insulation and holds everything together.

The heating element is wound up in a spiral for the stove-top burners. In the oven, there is one heating element around the bottom edge for baking, and one around the top edge for broiling. The baking element heats up the air in the oven, which cooks the food. The broiling element does that, and also browns the top of the food. Electric stoves control the temperature of their burners and oven with thermostats, which cut off the flow of electricity to the heating element when it gets too hot, and start the flow again when it gets too cool, just like in the iron on page 52.

An electric burner heats by first becoming red hot itself.

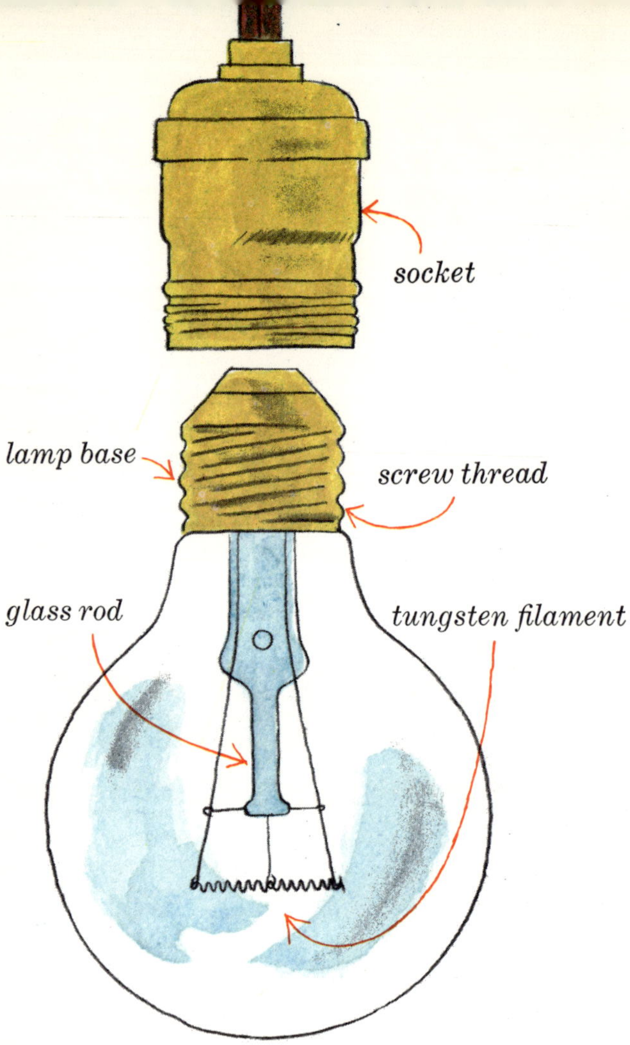

A wall switch is used to connect or disconnect the flow of electricity to the bulb. The wire leading into the bulb actually has two wires inside, one for the electricity flowing into the bulb and one for the electricity flowing out.

An ELECTRIC LIGHT BULB is a glass bulb with a threaded brass top that screws into a socket. Inside, a glass rod supports a frame of wires that hold up a very thin, coiled wire called a filament. It is made of tungsten, a special metal that can stand extreme heat. Before the bulb and its contents are sealed up tight, all the air is taken out and a special gas is put in. This is done because the filament would burn out very quickly if the oxygen of the air were present.

When the light bulb is screwed into a socket and the current is turned on, this is what happens inside the bulb. The electric current goes into the bulb and through one side of the wire frame to the filament. It flows through the filament, down the other side of the wire frame, and out of the bulb. The filament heats up when the electricity flows through it, just as the heating element of a stove does. But, at its hottest, the stove heating element glows only dull red. The light bulb filament gets much hotter than that, so hot that it glows white and gives off the strong light that you read and work by.

Thomas Alva Edison, the famous inventor of the phonograph and many other things, invented the light bulb. His hardest problem was finding a filament that would glow brightly and last a long time. He tried many materials before he found tungsten.

With the help of electric light, there is almost as much going on at night as there is during the day. Just think of playing night baseball before the light bulb was invented!

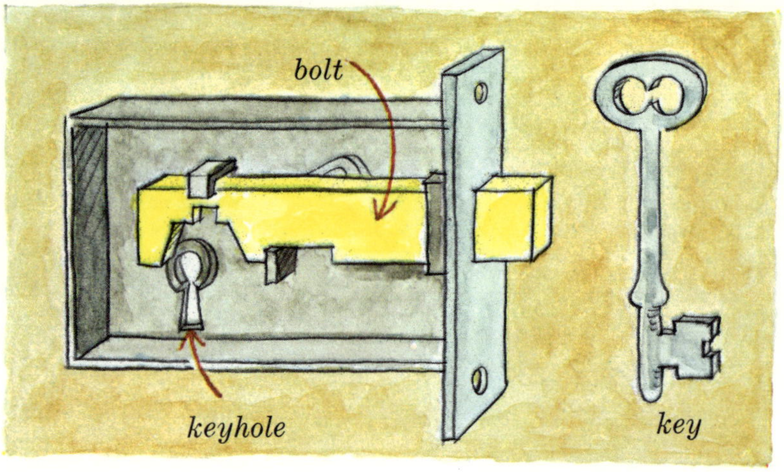

A SIMPLE LOCK uses a bolt that slides back and forth when the key turns. Each lock has a special key that will work it. The flat part at the end of the key has little notches cut into it in just the right places, so it can turn without being blocked by barriers, which are in the lock to make sure that only one key can work it. When the key turns, it catches against the bolt and pushes it into a hole in the door frame to lock the door. The key turns the other way to pull the bolt back and unlock the door.

A CYLINDER LOCK uses a key with grooves cut into its sides and little hills and valleys cut into its edge. The side grooves must match the shape of the keyhole. If they match, the key slides into the barrel (the part of the lock that turns) and runs under a series of little rods which rest on the hills and valleys. These rods are all of different lengths and are in sets of two, one on top of the

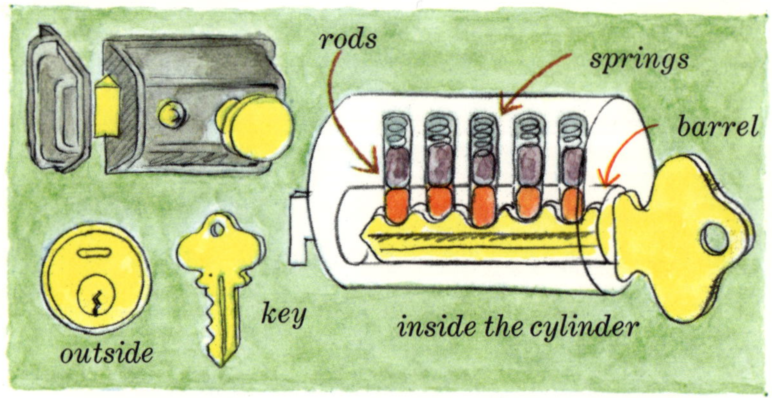

other. When the right key is in the lock, the hills and valleys will position the rods so that the tops of the lower rods line up with the top of the barrel. The key can then turn the barrel, moving the bolt aside.

A BANK VAULT uses a combination lock and a time lock. A combination lock needs no key. You turn the dial back and forth to the numbers of the combination, lining up the notches in three discs inside the lock. Then a release mechanism can pass through the notches, and you can pull the lever that draws back the bolts. The time lock sets a clock connected to the combination mechanism. Until the clock has turned to the preset time, the lock cannot be opened.

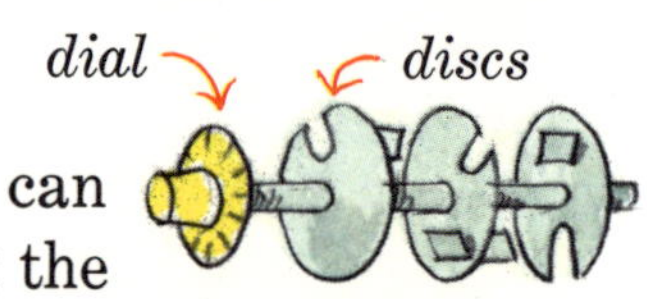

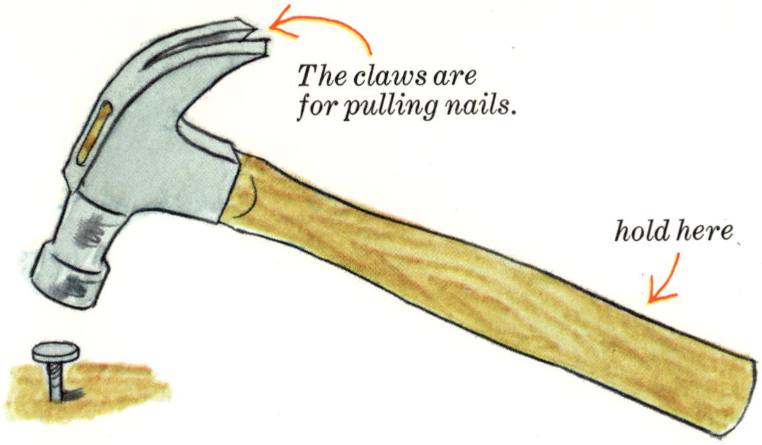

A **HAMMER** has a long wooden handle. When you hold the hammer handle at the bottom end and swing the metal hammerhead at a nail, your hand only moves a little, but the head, out at the other end of the handle, moves much farther and faster. So the head hits the nail with much greater force than it would if you held it in your hand and banged the nail with it. So, the long handle makes it easier to drive a nail. And that's why it's not so easy to pound a nail in with a rock.

A **SAW** has a wide steel blade with pointed teeth all along the cutting edge. If you look closely at the teeth you will see that each one is like a little chisel. As you push the saw forward each tooth chisels off a tiny piece of wood. The pieces fall to the floor as sawdust. The first tooth is bent to the left, the second to the right, the third to the left, and so on, so that the cut in the wood is wider than the saw blade. This allows the blade to be pushed back and forth easily.

A **DRILL** bit for wood has two cutting edges that wrap around its core like spiral staircases, and a pointed screw in the middle that goes into the wood first. This makes it easy to start the hole where you want it. Put the point of the screw where the center of the hole is to be and turn the handle, so the screw starts going down into the wood. After the screw part is in, the sharp points of the knife-like cutting edges cut into the wood, and start digging it out. The bits of cut wood move up out of the hole in the spiral space along the sides of the drill, like people going upstairs.

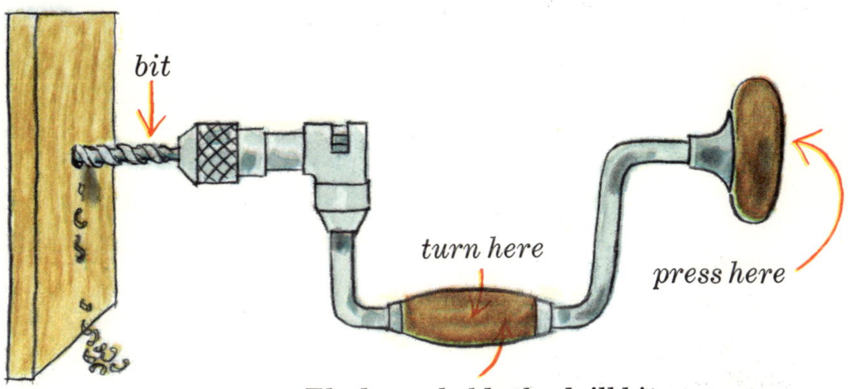

The brace holds the drill bit.

A **PLANE** is used to smooth out a piece of wood or to shave down something just a little to make it fit, like a door that won't close. You could do the same job with a knife, but the plane makes a much smoother cut. It has a wide, flat bottom with a wide, sharp blade sticking down through a slot in it. As the plane slides over the wood, the blade shaves off a layer. The more the blade sticks out, the more it cuts, but the harder it is to push. A plane has a handle in the back for pushing, and a knob in the front for holding it tight against the wood.

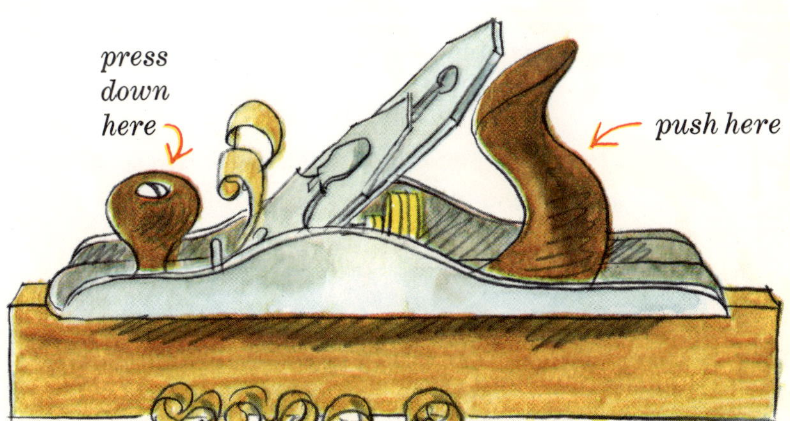

A simple TYPEWRITER has four important parts. 1) The keyboard has rows of keys, each marked with a letter, number, or symbol. 2) Each key is connected by levers to a metal type-bar, which has a raised metal edge in the shape of a letter, number, or symbol on its typing face. 3) The rubber roller holds the paper and is mounted on a carriage that moves along from right to left. 4) The inked ribbon runs in front of the roller.

When you strike a key, the levers move as the diagram shows, and the type hits the ribbon and leaves an imprint of the letter on the paper. This is why the machine is called a typewriter. As soon as the type-bar falls back from the ribbon, the carriage moves over one space. When you come to the end of a line you push the carriage return lever, moving the carriage back to its starting position and moving the paper up to begin a new line.

A **FOUNTAIN PEN** has a place to store ink inside it. The one in the picture has a rubber tube. To fill the pen you pull a little lever down. The lever squeezes most of the air out of the tube. Then you dip the pen into the ink in the bottle and release the lever. This lets the tube open and ink is drawn up into it, replacing the air you squeezed out.

When you write, the ink runs down the tube onto the point and goes onto the paper because of "capillary action." What is capillary action all about? Water clings to surfaces. When you hold a paper towel over some water and dip just the corner of the towel into the water, you will see the water actually climbing up the towel. In the same way, the ink sort of climbs out of the pen onto the paper.

A **BALLPOINT PEN** also stores its own ink in a plastic or metal tube. The tube is stopped up at the writing end by a tiny metal ball mounted in a socket. This ball rolls around as you write, picking up ink from the inside, and rolling it out onto the paper. The surface of the ball is slightly rough to help it pick up ink. The ball takes the place of the point used in the fountain pen, and that is why this pen is called a ballpoint. Ballpoint ink is not the flowing kind used in fountain pens, but a much thicker kind, more like mud than water. Otherwise, it would run out of the pen too fast.

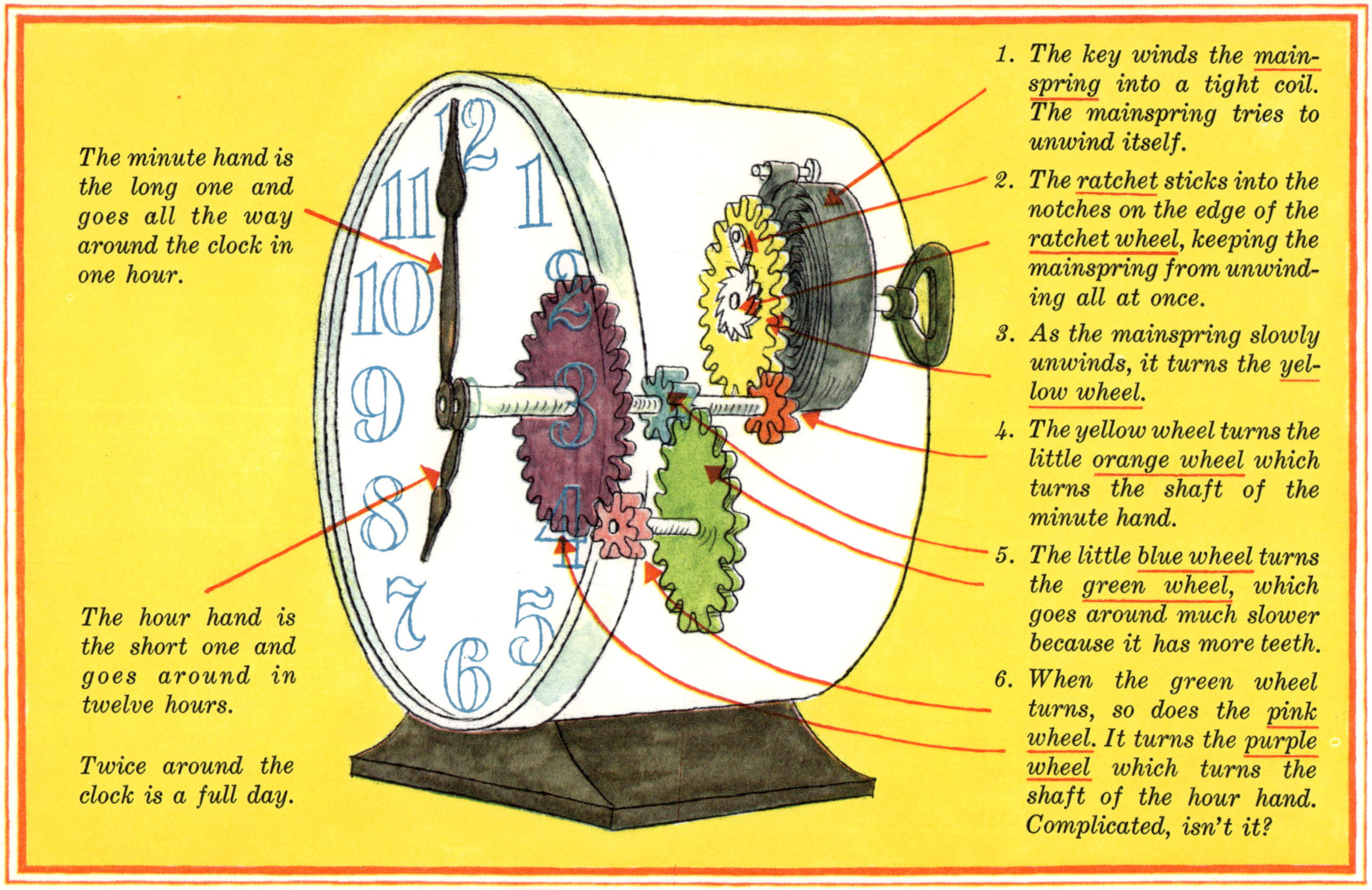

Any CLOCK needs power to make it run, and some kind of timing mechanism to control the rate at which it runs. It has to run at an exact rate so we can tell time by it. Some clocks use a weight for power and a swinging pendulum for timing. Electric clocks can use alternating current for power *and* timing, since the current alternates at a fixed rate.

A windup clock uses springs for power and timing. The picture above explains how the mainspring drives the hands. The hairspring, which controls the rate at which the hands are driven, is shown in the diagram below. It coils and uncoils just the right number of times per minute, moving the pallets in and out of the teeth of the escape wheel. This lets the escape wheel turn a tooth at a time. In this clock the green wheel (which you can see in both pictures) is where the power and the control come together. The mainspring turns the green wheel, but the escape wheel lets it turn just one tooth at a time, at the timekeeping rate set up by the coiling and uncoiling of the hairspring.

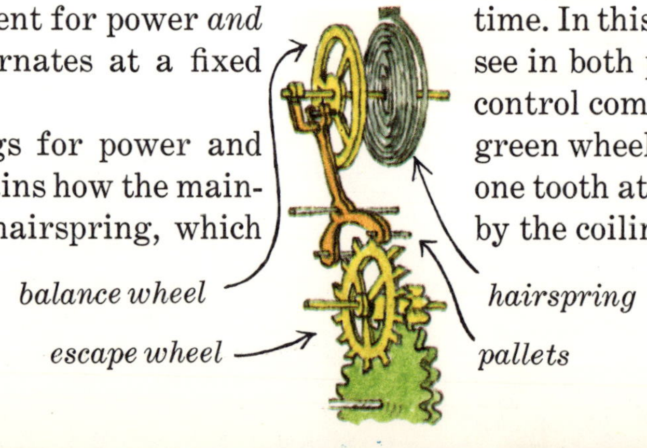

There are clocks that do more than tell what time it is. Some tell the day, month, and year. Some have figures and animals that play music, ring bells, dance. All work by clock machinery.

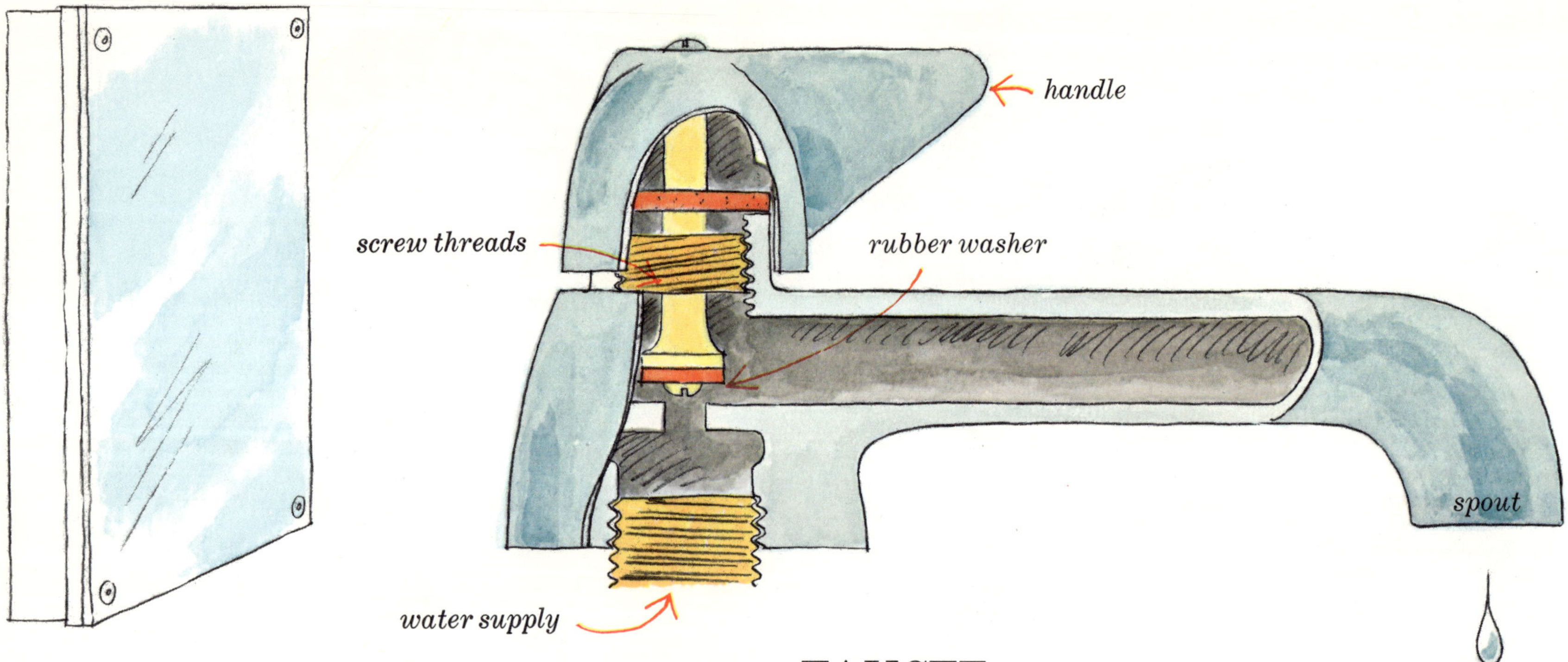

The FAUCET. The pipes in your house are always full of water, and you use the faucet to hold the water back or let it out. The inside of the faucet is divided in two by a wall with a hole in it. When the faucet is turned off, a rubber washer on the shaft of the faucet handle is pressed down over the hole. Water fills the area below the hole but it can't get past the washer into the spout. To turn on the faucet, you turn the handle. The handle shaft has screw threads on it and works like a screw. When you turn the faucet on, you are really unscrewing the handle a little. The shaft moves up, the washer is pulled away from the hole, and water flows through the hole and out the spout.

If you have a dripping faucet, the washer is probably worn. Water leaks around it and drips out the spout. All the faucet needs is a new washer.

The TOILET. When you flush the toilet, two things happen. The tank empties into the bowl and automatically fills up again. And the bowl empties into the drain pipe and automatically fills up again.

When you press the handle down, it pulls up the stopper which covers the pipe opening in the tank bottom. The tank water rushes through the pipe into the bowl to flush it. As the water in the tank goes down, so does the float. Its arm pulls open the inlet valve, and water starts coming into the tank. The stopper falls back over the pipe opening, the water rises in the tank, and the float goes up with it. When the tank is filled to the right level, the float has risen high enough to close the valve.

In the meantime, the tank water pours into the bowl, and the water level in the bowl goes up. This pushes the water up and over the curve in the drain-pipe. As the water runs down the pipe, the water and waste in the bowl are sucked along behind it, until the level gets so low that air can get into the pipe. Then air goes down the pipe instead, and the bowl fills up again.

If your toilet won't stop running, here's what might be wrong. The stopper may not be falling back over the pipe opening, so the water trying to fill the tank keeps running into the bowl. Or the float arm may not rise high enough to shut off the inlet valve, so the water just keeps coming in and running out the overflow pipe. You and your mom or dad can fix this easily, by adjusting the parts.

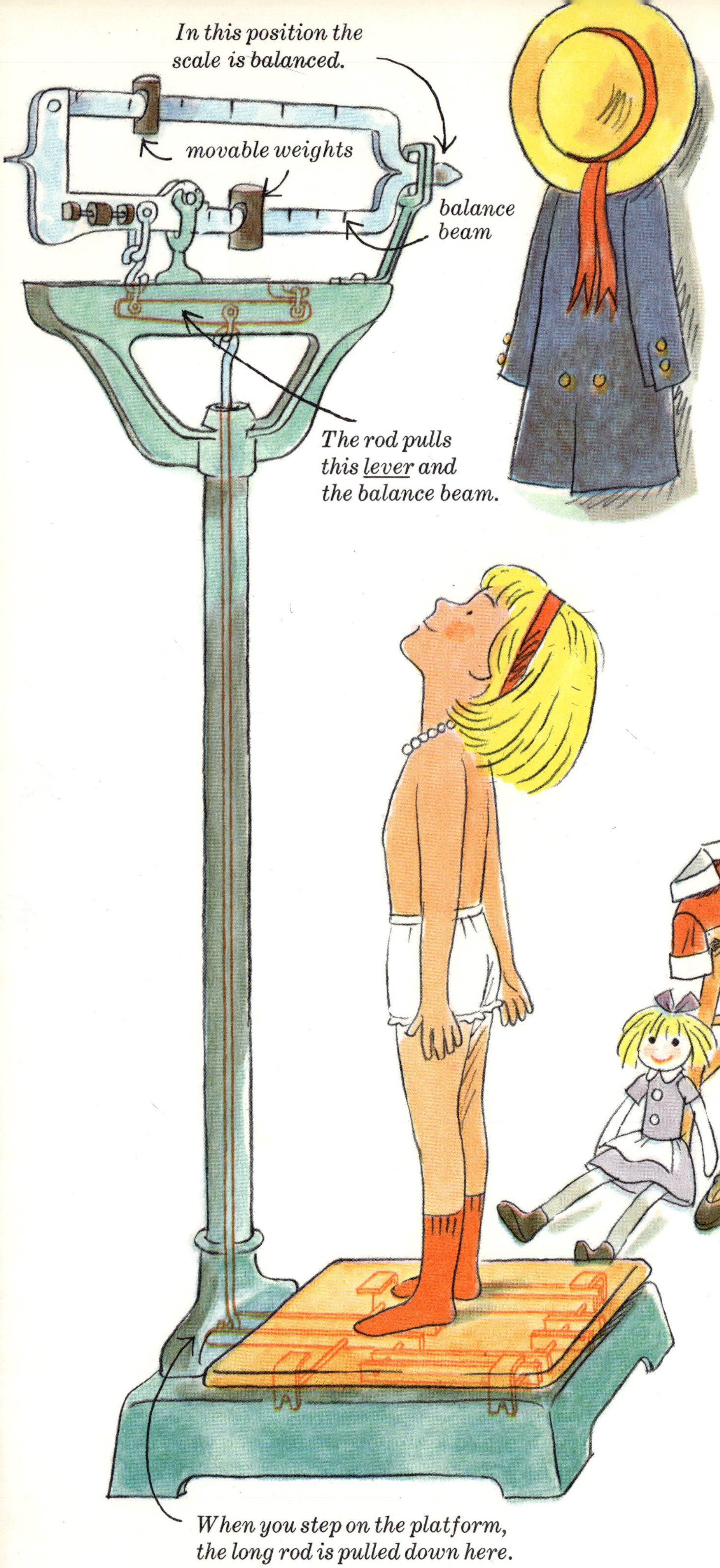

A SCALE measures your weight and helps the doctor tell how fast you are growing.

The scale is like a see-saw. You could actually use a see-saw to compare your weight to a friend's. You sit on one end, and the other end goes up. Now your friend gets onto the see-saw at the center, and moves out toward the other end until it balances. Where he is when it balances tells you if you weigh less than, more than, or the same as he does. With a real scale, you can tell exactly how much you weigh, using the weight scale marked on the balance beam.

When you stand on the scale's platform, your weight pushes down a lever underneath. The lever acts on a series of rods and levers which pull one end of the balance beam down. The other end goes up, like the other end of the see-saw. The doctor moves the weights out toward the end until the beam balances (just like your friend moved out on the see-saw). The numbers next to the weights tell how much you weigh. The scale is adjusted (calibrated) so that when the weights line up with the numbers that tell your actual weight, they exactly balance the pull of your weight through the rods and levers.

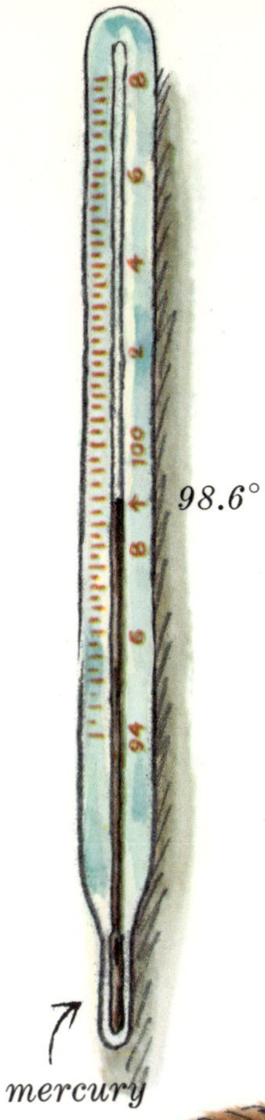

A THERMOMETER measures your temperature—how warm you are. Warmth is measured in degrees, and if your temperature is 98.6 degrees it's just about right. If it goes much higher, you are too warm—you have a fever. A thermometer is made of glass, with a hollow bulb at the bottom, and a very thin tube inside leading from the bulb to the top. The bulb is filled with mercury.

Like the thermostat (page 52), the thermometer works because things expand (get bigger) when heated. The special idea of the thermometer is that the bulb at the bottom holds a lot of mercury. When the mercury expands even a little, it moves a long way up the thin, hollow tube, so just the smallest difference in temperature will move the mercury far enough in the tube for your eye to see the change.

When you take your temperature, the warmth of your mouth makes the mercury expand and go up the tube. Lines and numbers mark the degrees. The number next to the top of the mercury is your temperature.

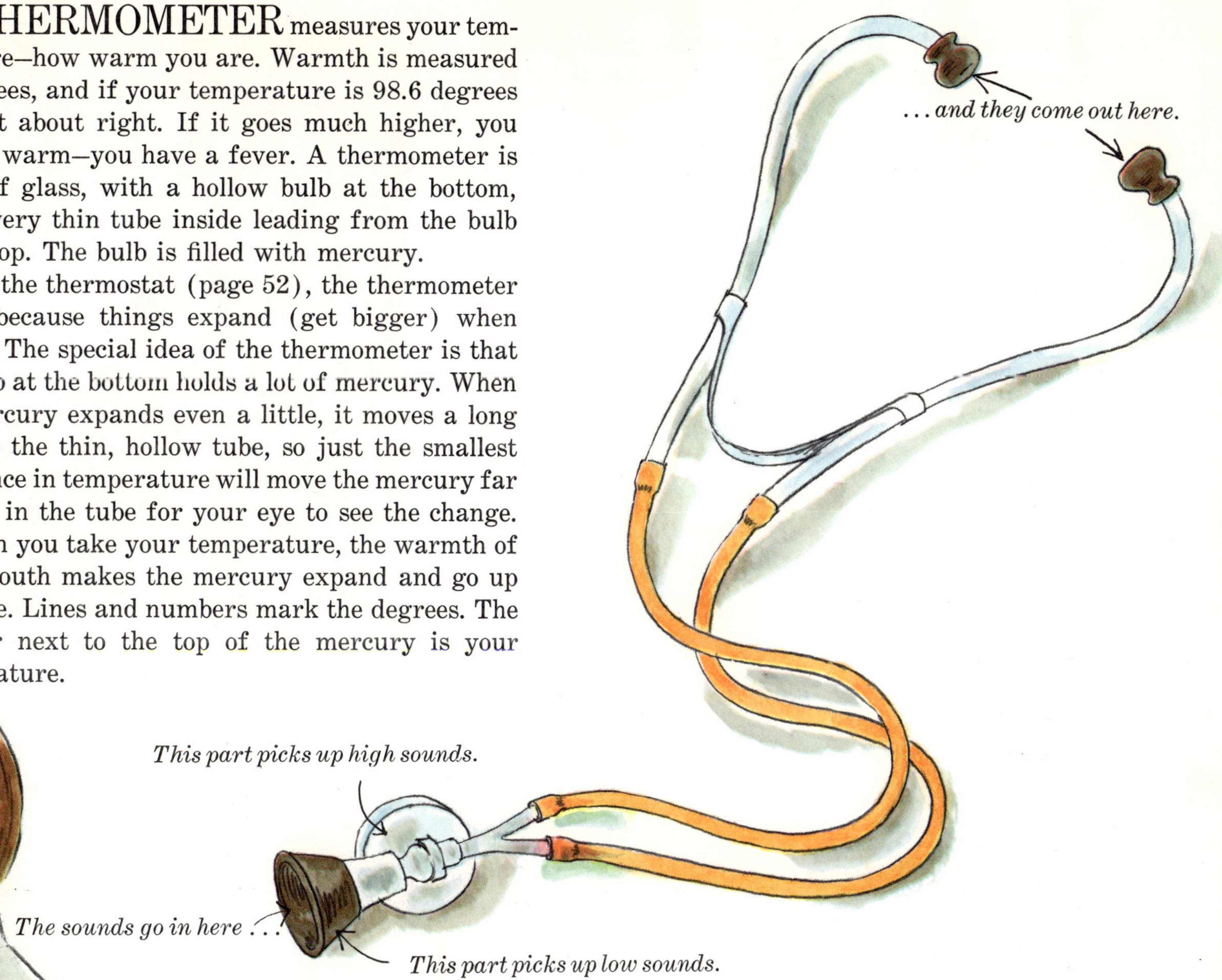

A STETHOSCOPE helps the doctor hear the sounds your body makes inside. It is like a direct line from your body to the doctor's ears. The doctor presses the wide end firmly against your body, to keep any noises from outside out of the stethoscope. The wide end gathers the body sounds, and sends them up the long rubber tubes, through the earpieces, into the doctor's ears. The earpieces fit snugly into his ears so that only sounds coming through the stethoscope can get in. The doctor hears only what's going on in your body—your heart beating, your lungs breathing, your stomach digesting. All these sounds tell him how the things that he can't see are working.

A PIANO. All sounds that you hear are really vibrations of the air that travel like waves to your ears. The more rapid the vibrations, the higher the sound. In one family of musical instruments, strings do the vibrating. The shorter, thinner, or tighter these strings are, the faster they vibrate and the higher they sound.

A piano makes its musical notes by the striking of strings with little hammers, which work when you press the keys. The strings, which are really metal wires, are stretched across a metal frame.

The diagram shows the way one key works. When the key is pressed down, the damper rises to let the string vibrate, and the hammer strikes the string and falls back. As long as the damper stays up, the string keeps on vibrating and making sound. When you release the key, the damper falls back on the string, stopping the sound.

The vibrating strings would not make much of a musical sound without the sounding board, a thin sheet of wood underneath them. It vibrates, too, making the sound louder and fuller.

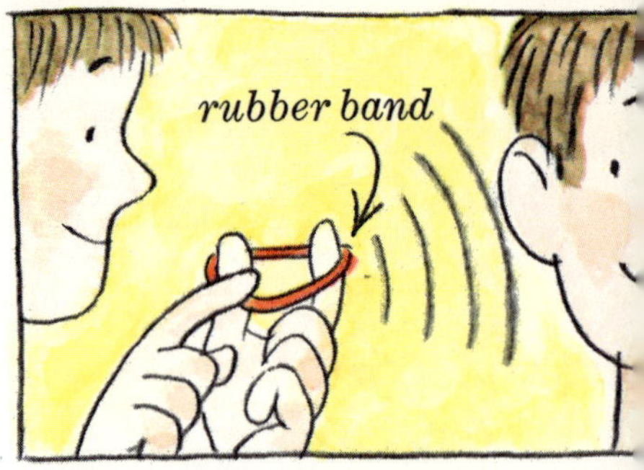

All sounds are made by something vibrating. When you pluck a rubber band, it vibrates and makes sound waves in the air.

The keyboard has eighty-eight keys. Some are black and some white. The color and shape of the keys helps the pianist to find the ones he wants to play.

what happens when you press a piano key

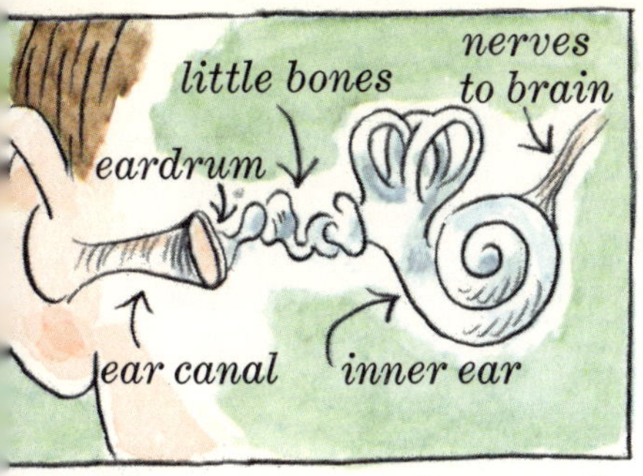

Inside your ear, the sound waves hit the eardrum. It vibrates and moves the little bones behind it. The bones move a liquid in the inner ear, which stimulates nerves to send messages to the brain.

A VIOLIN has four strings, all the same length but of different thicknesses. The thickest vibrates slowly and plays the lowest notes. The thinnest vibrates fast and plays the highest notes.

The strings run from the tuning pegs over a wooden bridge, and are fastened to the tailpiece. Because they touch the bridge, their vibrations go down through it into the body of the violin, which makes the sound louder and fuller.

When the violinist draws the bow across the strings, the horsehair rubs against them and starts them vibrating. To make a higher note, he presses the string down onto the fingerboard with his finger to make the vibrating part shorter.

The forty-seven strings are color-coded so the harpist can easily find the ones he wants.

A HARP has forty-seven strings and seven pedals. The harpist plucks the strings with both hands, and presses the pedals with his feet to tighten the strings and make them sound a half or whole tone higher. Just like in a piano, there is a sounding board which vibrates with the strings and amplifies the sound.

The long thick strings at the front of the harp vibrate slowly and play the low notes. The shorter and thinner strings toward the back vibrate faster and play the higher notes.

A TUBA. Most wind instruments make sounds by making a column of air inside them vibrate. The shorter the column is, the faster it vibrates, and the higher it sounds. When you play a bottle by blowing across the top, the air inside vibrates. If you add water, the air column gets shorter and the note gets higher. Wind instruments have special systems for changing the length of their air columns.

The tuba is a long tube all coiled up. It makes low notes because its air column is so long. When you blow into the mouthpiece, your lips vibrate and set the air vibrating. Sound comes out. To change the note, you change the length of the air column by pressing the valves. You can also change the note by changing the tightness of your lips.

A CLARINET is a long, wooden tube with a mouthpiece at one end and a wider bell at the other. To set its column of air vibrating, it uses a reed, a thin piece of cane clamped to the mouthpiece, instead of the player's lips. When the clarinetist blows into the mouthpiece, the reed vibrates, and starts the air in the clarinet vibrating, making a note.

As he plays, the clarinetist covers some of the holes along the tube with his fingers. Air escapes from the rest of the holes, and this determines how long the vibrating column of air is, and so, what note is played. Just as for the tuba, a long column of air vibrates slowly, making a low note, and a short column vibrates faster, making a higher note.

A HARMONICA uses many reeds of different lengths, one for each note it can play. The long reeds play the low notes, and the shorter ones play the high notes. Each slot has two reeds, one on the top and one on the bottom. When you blow into the slot, the top reed vibrates and plays a note. When you draw air in through the slot, the bottom reed vibrates and plays a different note.

Take a small can like a soup can, with the top off. *Make a little hole in the bottom.* *Put a string throug*

1. When you speak, your voice will make the bottom of this can vibrate. If you hold the string taut, the vibrations travel along it.

The TELEPHONE works like the tin can and string phone, but, of course, much better. The tin can and string phone sends the sounds as vibrations, while the real telephone sends the sounds as electrical signals that can travel long distances in wires. When you speak, sound waves go into the transmitter of your telephone, change into electrical signals which go to the other receiver, and there are changed back again to sound.

This is the way it works. Your vocal cords,

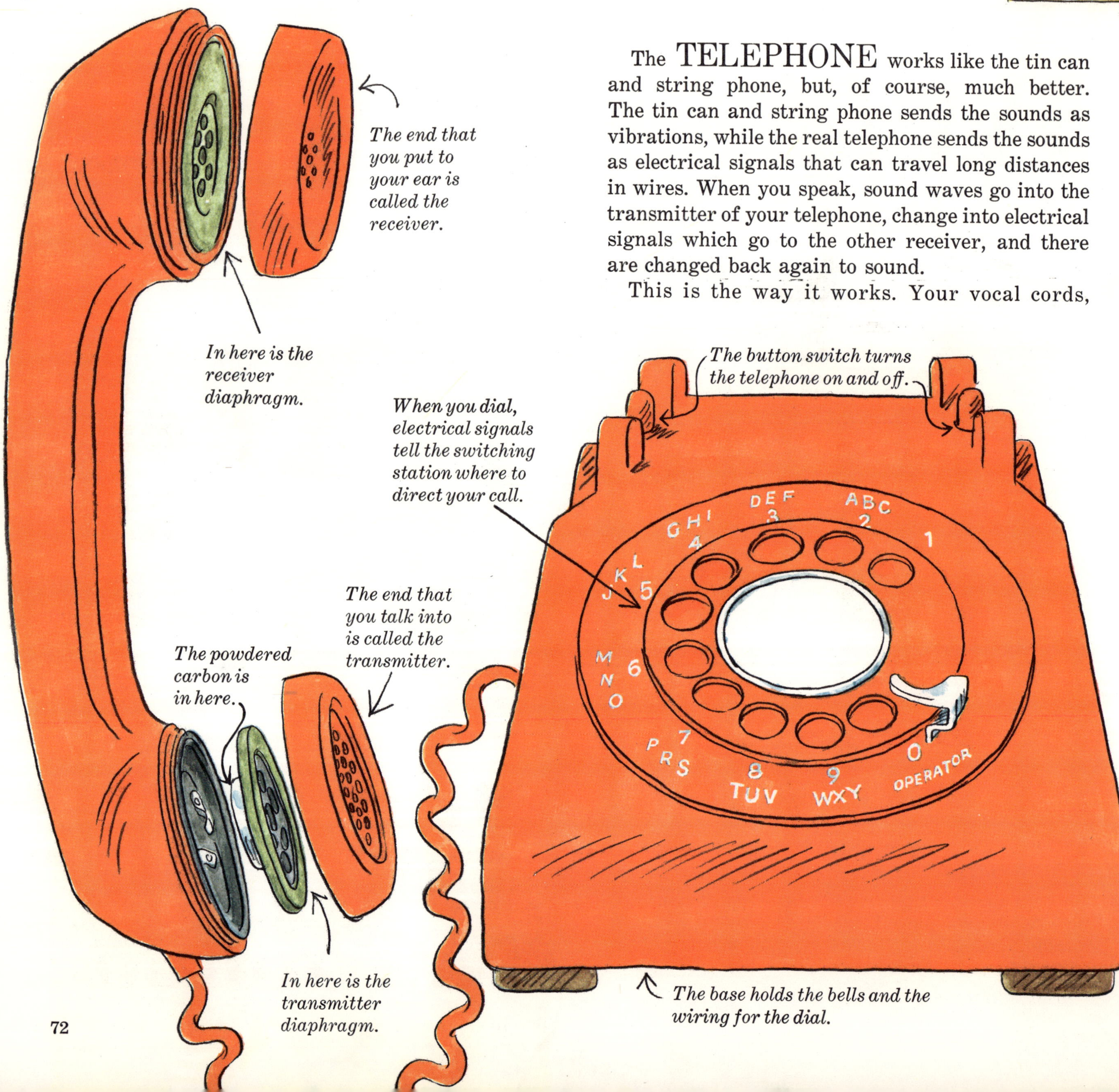

The end that you put to your ear is called the receiver.

In here is the receiver diaphragm.

The button switch turns the telephone on and off.

When you dial, electrical signals tell the switching station where to direct your call.

The end that you talk into is called the transmitter.

The powdered carbon is in here.

In here is the transmitter diaphragm.

The base holds the bells and the wiring for the dial.

and tie a knot inside. *Do the same thing at the other end of the string.* *Now you have a simple telephone.*

2. The vibrations traveling along the string make the bottom of this can vibrate, and it makes sounds similar to your voice.

tongue, and lips form sounds which are really pressure waves in the air. These waves strike a flat, round piece of thin metal, called a diaphragm, inside the transmitter and make it vibrate back and forth. On the other side of the diaphragm there are grains of carbon which are squeezed together and then let loose as the diaphragm vibrates back and forth. When the carbon is squeezed together, lots of electricity can pass through and when it is loose less current can pass through. The current varies just the way the sound waves striking the diaphragm vary. This current goes along the wire that finally leads to the receiver of the other telephone where someone is listening. In the receiver is another metal diaphragm with a small electromagnet behind it. As the changing electric current goes through it, the electromagnet pulls the diaphragm back and forth, matching the vibrations at the speaking end of the first telephone. The vibrating diaphragm sends sound waves into the listener's ear.

It is possible to talk by telephone to people in most parts of the world (but you need to know each other's language).

On land, the wires going from one telephone to another are strung up on poles or buried underground. Cables on the ocean bottom connect one country with another.

Now some telephones let you see the person to whom you are talking on a small television screen.

BROADCASTING. The voice of a disc jockey, the music he plays from the radio station, and the pictures you get on TV, are all sent to you through the air. We can do this because radio waves (which scientists call electromagnetic waves) travel through the air with the speed of light and can be made to carry these kinds of information. They don't carry it in sound or picture form, though, but as electrical signals which are a sort of "code" for the sounds or pictures. The radio or TV broadcasting station changes the sound or pictures into the code. You need a radio or TV to do the decoding.

The transmitter the disc jockey works with sends a powerful carrier wave out into the air from a broadcasting antenna, which is usually a tall tower. The waves spread out from the top, the way ripples of water spread out when you throw a stone into a pond. Each station's carrier wave is different from every other station's. It has a different wavelength (the distance between the waves). This wavelength is used in tuning in one station or another.

The disc jockey's microphone works like a telephone transmitter (page 72), changing his voice to a pattern of electrical signals which are combined with the carrier wave as it goes into the air. The waves can travel hundreds of miles before they get too weak to be picked up by another antenna. There are radio waves all around us in the air, but we can't hear them. We need a radio to decode them first.

Meet Guglielmo Marconi, father of radio. He invented a way of sending messages through the air without wires.

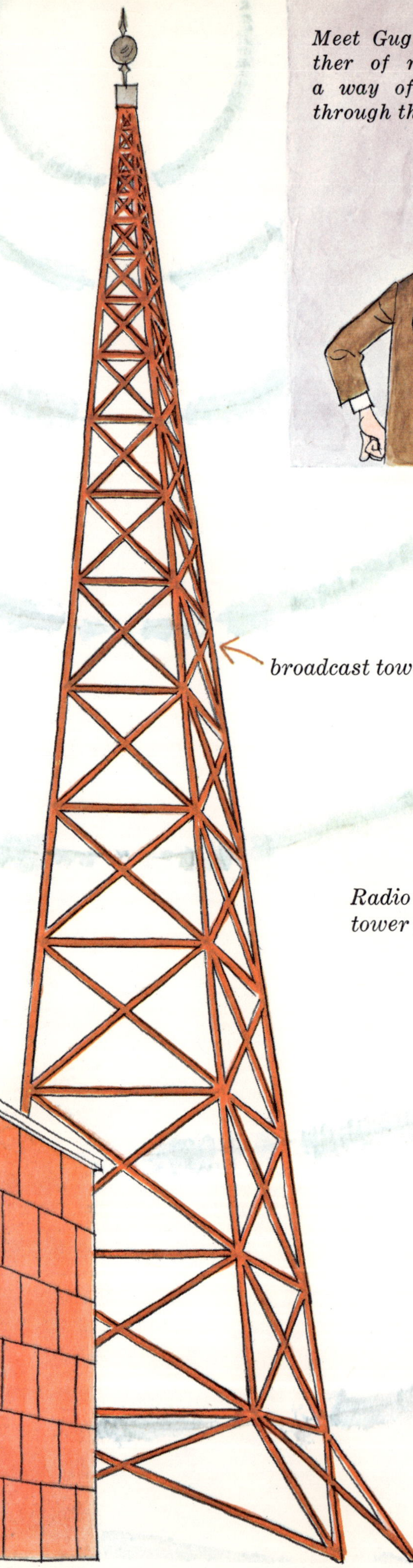

broadcast tower

Radio waves go out from the tower in wider and wider circles.

The microphone turns sound into electrical signals.

Most radios now have transistors to do all the things the tubes used to do. Transistors are much smaller than tubes, they last longer, and they warm up right away.

When radios looked like this . . . people looked like this.

The RADIO has an antenna, too, usually a coil of wire. Antennas can either send or receive, and this one receives radio waves from the air. Weak electrical signals, just like the ones in the studio, are created in the antenna wire. The antenna, however, does not pick up just the radio waves you want. It picks up any that come to it. So the electric current going into the set is a combination of many different signals. It goes through the tuner, which selects out the carrier wavelength of the station to which you are tuned. Then the sound signal is separated from the carrier wave. It is so weak that it has to be made stronger (amplified) by tubes or transistors. (When you turn the volume control knob to make the music louder, you are actually adjusting the amplifier.) Finally, the amplified current is fed to the loudspeaker.

The loudspeaker works very much like a telephone receiver. It is usually a cone of paper, which acts as a diaphragm. The current of sound signals runs through an electromagnet attached to the back of the diaphragm. As its strength varies with the signals, the electromagnet pulls the diaphragm back and forth. The vibrating diaphragm sends out waves to your ears, and you hear the radio program being broadcast. And all of this, from the studio microphone to your radio, takes just the smallest bit of a second!

1. Transistors amplify the sound.

2. The sound is recorded on tape.

3. The cutting machine cuts a wavy spiral groove into a plastic master disc.

4. The enlarged grooves look like this.

5. A metal mold is made from each side of the master. Warm plastic is placed between them, and they press together to make a record.

MAKING A RECORD. Sounds are carried by air. In the past, once the sound waves died out, that particular sound was gone forever. If you wanted to hear something, you had to be listening at the same moment the sound was being made. But, with the invention of the record player and the tape recorder, people could not only carry music around with them, captured on a record or tape, but they could hear it over and over again, when and where they wanted.

In the recording studio, the microphone changes the music to electrical signals which are amplified, and then recorded on tape. The musicians and studio technicians listen to the tapes. They adjust the balance of the instruments and voices, and sometimes add more music to the tape.

Finally, the master tape is played into the record cutting machine. The vibrations of the signals make the needle vibrate from side to side. As the plastic disc turns, the vibrating needle cuts a wavy spiral groove into it from the outside in. This groove is a physical "record" of the music.

Metal molds made from the master disc press out the plastic record that you buy in a store.

The RECORD PLAYER was invented by Thomas Edison back in 1877. Then, a record was a fat cylinder that you cranked around by hand. It made a funny, scratchy sound. Today, even a small record player can play music that sounds so clear it almost seems that you are hearing a live performance.

When you switch the record player on, the turntable starts turning at just the right speed, the same speed the turntable on the record cutting machine turned when the record was made. A needle at the end of the tone arm follows the sound grooves and vibrates. The vibrations set up weak electrical signals in the cartridge, and they go through the amplifier which makes them stronger. Now the loudspeaker, which works with an electromagnet like the radio loudspeaker, turns the strong electrical signals into waves of sound, and you can start dancing.

The tone arm guides the needle over the record, and presses it into the grooves.

The loudspeaker is under here.

An electric motor under the turntable turns it around at the speed you select, usually 33⅓ turns every minute.

speed selector

cartridge

The volume knob makes the sound loud or soft.

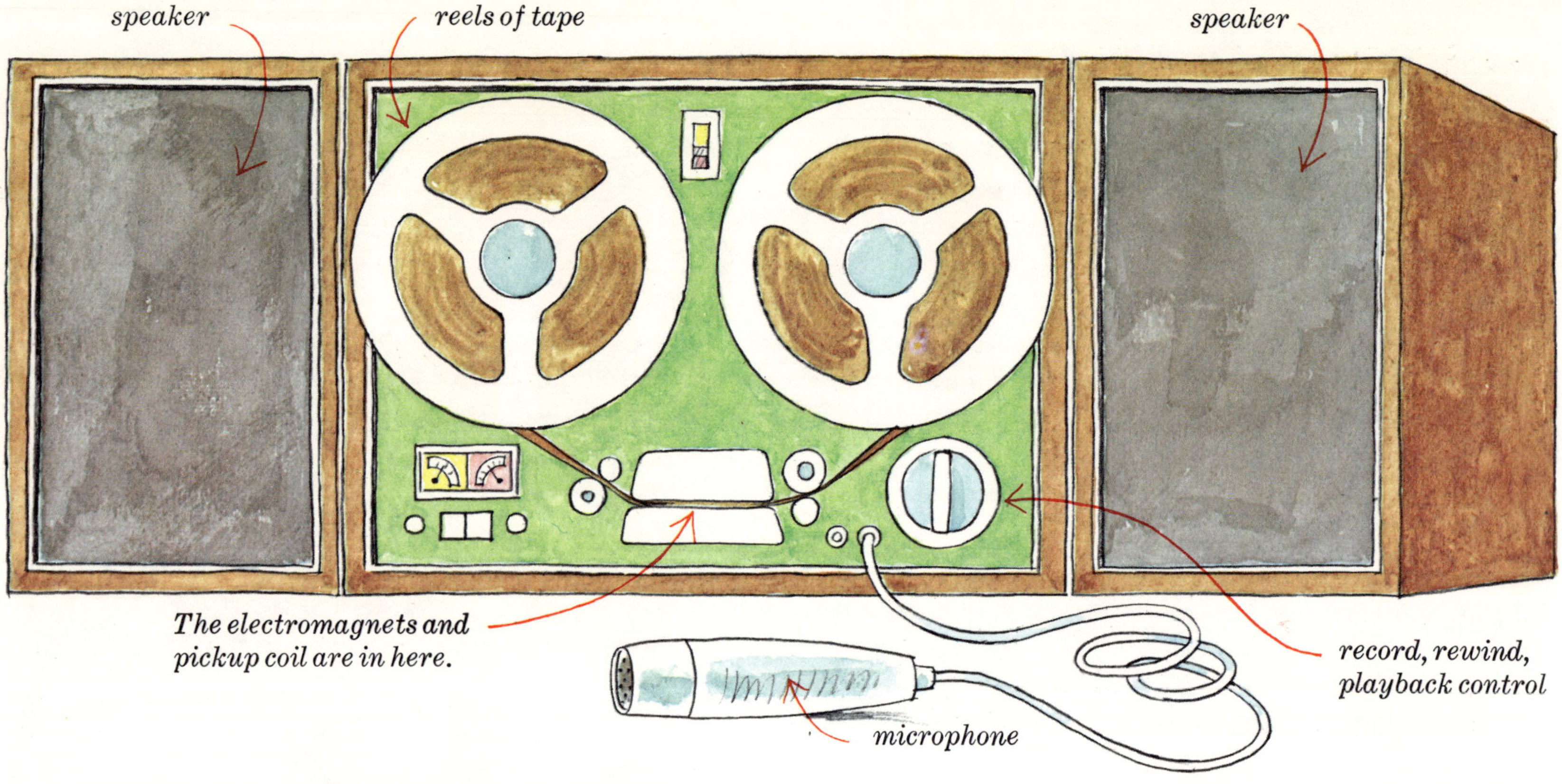

A TAPE RECORDER captures sound on reels of narrow plastic tape. To record, put a reel of tape on one of the two spindles and an empty reel on the other. Thread the tape through the right slots and fasten it to the empty reel.

When you press the "record" button, the motor inside rolls the tape slowly from the full reel onto the empty one. Now speak into the microphone.

The tape is coated with tiny iron oxide particles. During recording, the particles are magnetized in certain directions, creating a magnetic pattern on the tape that represents the original sound pattern.

Here is what happens. The tape rolls past two electromagnets (page 36 explains electromagnets). As the tape goes through the magnetic field around the first electromagnet, the magnetization of the little particles of iron oxide is removed, so that if anything had been recorded on the tape it is erased. Then the tape passes the second electromagnet, the one that records new sounds. This electromagnet receives the electrical signals you make as you talk into the microphone. Its magnetic field varies as the signals vary. As the tape passes by, the particles of iron oxide are magnetized depending on the strength of the magnetic field at that instant.

Now the sound has been recorded and you want to hear it. You press the "rewind" button and the tape winds back onto the first spool. Then press the "play" button and the tape starts through again. It passes the coil of a third electromagnet, and the magnetism of the metal particles on the tape reproduces the original current of electrical signals in the coil. The current goes into the speakers, and your voice comes out.

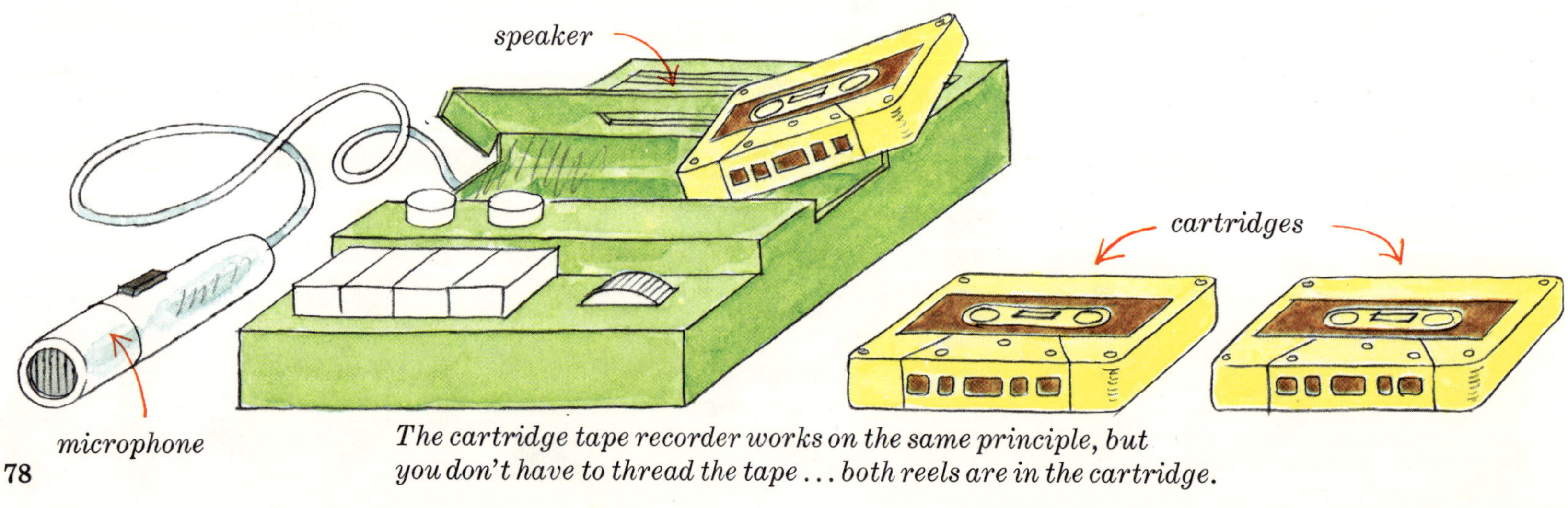

The cartridge tape recorder works on the same principle, but you don't have to thread the tape . . . both reels are in the cartridge.

Portable recorders are used by radio newsmen to record interviews and political speeches.

This is a good way to study foreign languages . . . talk into the microphone, then listen to yourself.

Recording baby's first words . . . wouldn't you like to hear what you used to sound like?

A tape recorder is useful to international spies . . . for recording important diplomatic secrets.

Once mirrors were made of polished metal. They scratched very easily. This one is Egyptian.

Remember that what you see in the mirror is backwards. What looks like your right ear is really your left ear.

A MIRROR is a smooth surface that reflects light. It can be made of silver, steel, bronze, glass, or anything smooth and shiny. The mirror we use most today is made of a sheet of glass with a very thin coat of silver on the back. The silver does the reflecting. The glass makes a very smooth transparent surface for the silver to coat, and protects it in front. A coat of dark paint protects it in back.

When you stand in front of a mirror, light rays travel from you to the mirror. They bounce off the shiny mirror, and come back to you, carrying an image of yourself. This image that you see reflected in the mirror is always backwards. Also, if you stand two feet away from the mirror, your image appears to be four feet away from you, as though it were behind the mirror. This is because the light rays travel four feet before you see them—two feet to the mirror, and two feet back to your eyes.

About 400 years ago, the mirror makers of Venice (the city that has canals instead of streets) began to make such good glass mirrors that people stopped making metal ones. The mirrors were very expensive, and how they were made was kept very secret.

mirror bent for chubby midget look

These curved amusement park mirrors change the way you look by reflecting the light rays in all different directions, instead of straight out, like a flat mirror.

mirror bent for that skinny giant look

In a KALEIDOSCOPE, you see a full circle, but you are actually looking at a pie-shaped wedge! It's all done with mirrors. Two mirrors set at an angle to each other run the length of the tube. At the far end of the tube are two glass discs with many colorful glass chips between them. When you hold the kaleidoscope up to the light, the mirrors reflect the colored chips, and instead of seeing one pie-shaped wedge, you see a circle made up of a repeating wedge-shaped pattern.

Turn the kaleidoscope and the designs keep changing as the chips fall into new positions.

LENSES are used to make things look larger or smaller. They do this by bringing together or spreading out light rays that go through them.

Lenses are transparent, and most are made of glass. When a light ray passes through a lens, it bends, according to the shape and curve of the lens. A lens that makes an object look larger spreads out the rays of light coming from the object so that they are wider apart when they get to the eye than they would be without the lens. So, the object looks larger, and you see a lot more detail in it than you would otherwise see.

Lenses are used in eyeglasses, cameras, projectors, anything involving the transfer of light rays.

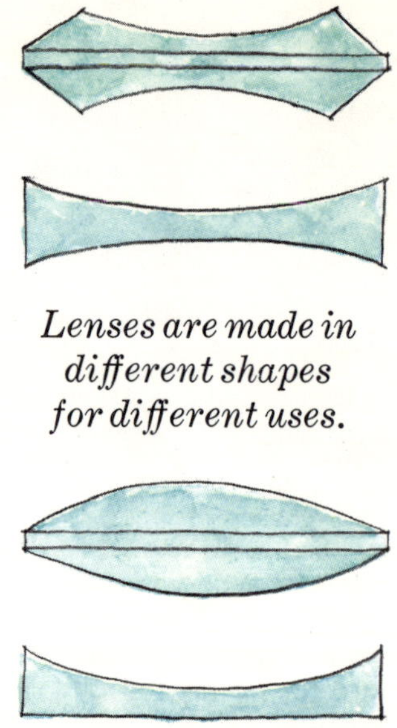

Lenses are made in different shapes for different uses.

magnifying glass lens

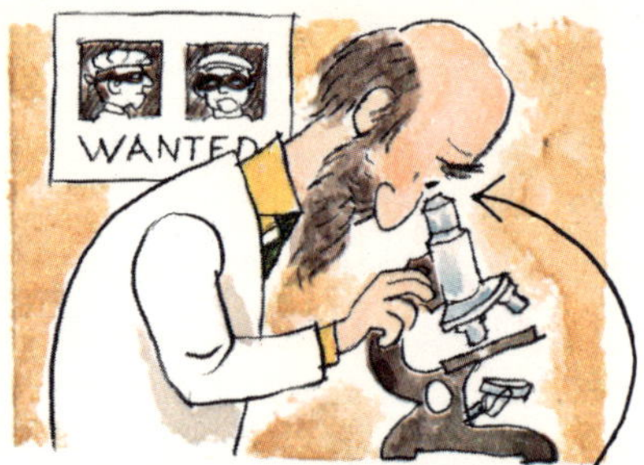

microscope lenses

The MICROSCOPE helps you see things that are too small to be seen by your eye or with the single lens of a magnifying glass. It has a series of magnifying lenses in a hollow tube. The lens nearest the thing you want to see magnifies it, and then another lens magnifies the image from the first lens. A third lens magnifies the image again. With such a microscope, scientists can study plant or animal cells, germs, almost anything they want to see.

To use the microscope, you mount what you want to look at (the specimen) on a slide. The specimen must be thin enough to be transparent, because light must go through it from underneath. Place the slide on the stage, with the specimen over the hole. The mirror under the stage reflects light up through the hole, the specimen, and the lenses in the tube, to your eye, where the specimen appears much larger and you can see all the tiny details.

Some microscopes have several different-strength lenses, mounted on a disc. You move the lens you need into line with the tube by turning the disc.

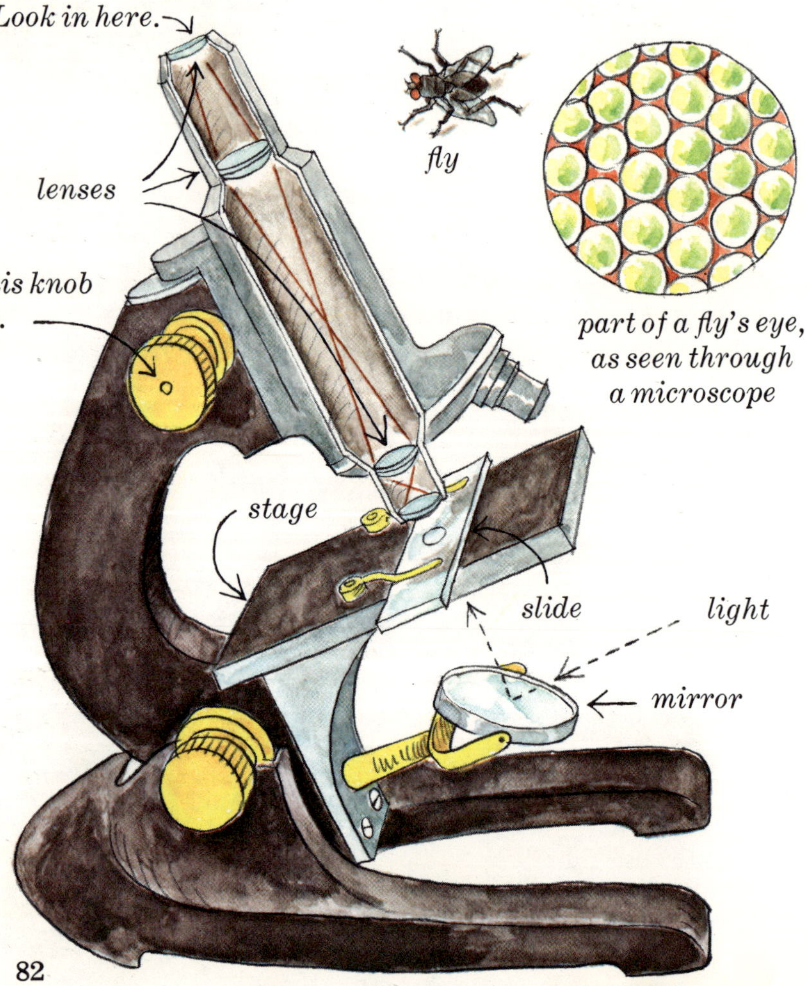

fly

part of a fly's eye, as seen through a microscope

*It's the distance that light travels through space in one year—almost 6,000,000,000,000 (6 trillion) miles.

A TELESCOPE also has a series of lenses mounted in a tube, but it is not for seeing tiny things that are close to you. It is for seeing big things that look tiny to your eyes because they are so far away. Astronomers use giant telescopes like the one in the picture to look at faraway planets and stars.

The telescope was invented by a man named Galileo 300 years ago. The first telescope looked like this.

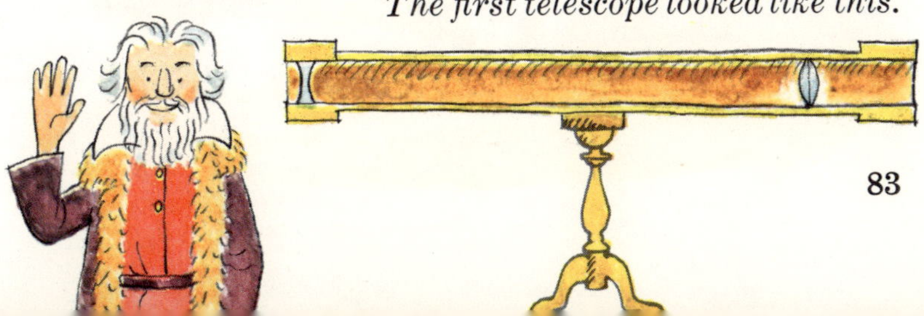

A 35 millimeter camera uses film 35 millimeters wide. The bottom opens for loading, but the top is off here, to show the insides.

The CAMERA uses light rays to record an image on film. The rays bounce off the subject of your picture into the lens, which focuses the rays on the film at the back of the camera. The lens must be the right distance from the film to give a sharp, clear picture, and getting this just right is called focusing the camera.

For the best pictures just the right amount of light must hit the film. You can adjust this by making the lens opening larger to let in lots of light, or smaller to let in less light.

You can also change what is called the shutter speed to adjust the amount of light. Behind the lens opening in this camera, the shutter curtain lies across the film, keeping light out. It has a very narrow up and down slit in it at one side. When you press the shutter button, the slit travels across in front of the film, letting in light across the picture area. When you adjust the shutter speed, it changes the width of the slit. When the slit is wide, it lets in a lot of light; when it is narrow, it lets in only a little. In other cameras, the shutter is in the lens, and is made of five metal leaves that open out from the center, and then close together again.

People used to have to hold still for minutes to have their pictures taken. Now we can snap a picture in a thousandth of a second.

FILM has a chemical coating. Where light rays hit it, chemical changes occur. Other chemicals are used to develop the film, producing either a negative or a positive transparency.

1. A black and white print is made by shining light through the white and black negative onto chemically coated paper, which is then developed.

2. Color prints come from color negatives. Color film has a different kind of coating from black and white film.

3. Color slides are transparent like negatives, but they are positives. They need light shining through them to be seen.

A **SLIDE PROJECTOR** is used to show slides on a screen. It shines light through a slide, and the light rays carry the image to the screen. In this automatic projector the slide tray sits on top. For viewing, a slide drops into a slot behind the lens. Each time you press the button, the slide you've just seen pops back up into the tray, and the next slide goes down into the slot.

In the back of the projector is a light bulb. A mirror behind it makes sure that almost all the rays from the bulb go forward. The condensing lenses in front of the bulb gather as many of the rays as possible and spread them evenly over the slide. The rays go through the slide, through the focusing lens, and onto the screen. The focusing lens can be moved in or out to make the picture sharp and clear.

The lamp in the projector gets so hot it could damage the slides. A thick sheet of glass in the condenser absorbs a lot of the heat before it reaches the slide. There is also a little fan in the projector, which pulls cool room air in and around the slide.

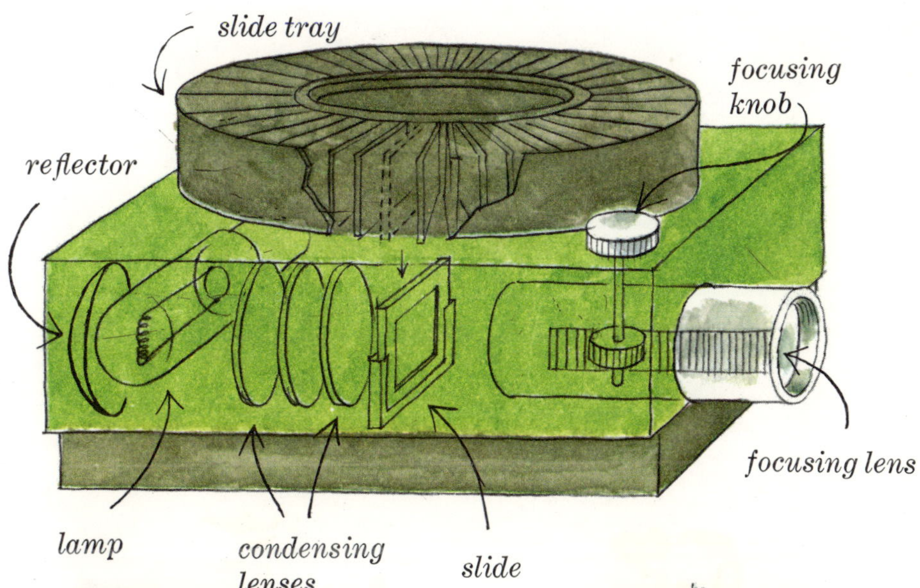

A MOVIE CAMERA takes pictures just like a still camera, except that it takes a lot of pictures of a moving object, one right after the other, so that when they are projected on the screen, they seem to move. Each of these pictures is called a frame. Here is how the camera works.

A spool of film is put into the camera, and the film is threaded past the lens, to the take-up spool. When you start the camera, a battery drives a small motor which turns the sprocket wheels. These have teeth that fit into the holes on the edge of the film and move the film through the camera. The pressure plate holds the film flat as it goes past the lens, a frame at a time.

After each frame has been exposed, a small claw fits into one of the holes and pulls the film ahead just enough so the next frame is behind the lens. The claw moves the film past the lens in jumps the size of one frame, while the sprocket wheels pull the film smoothly from the feed spool, to the lens, to the take-up spool. There is a loop of film on each side of the lens so the actions of the claw and the sprocket wheels won't interfere with each other.

Light must not hit the film while it is being pulled from one frame to the next or the picture will be blurred. So there is a shutter that covers the film for that moment. It is a metal disc that spins around between the film and the lens. An opening in the disc spins past the film to let the light hit it. Then the solid part of the disc covers the film while the claw advances it. Then the opening spins by again to expose the new frame. This spinning shutter is called a rotary disc shutter.

Movie of a smile . . . watch the smile get bigger and bigger.

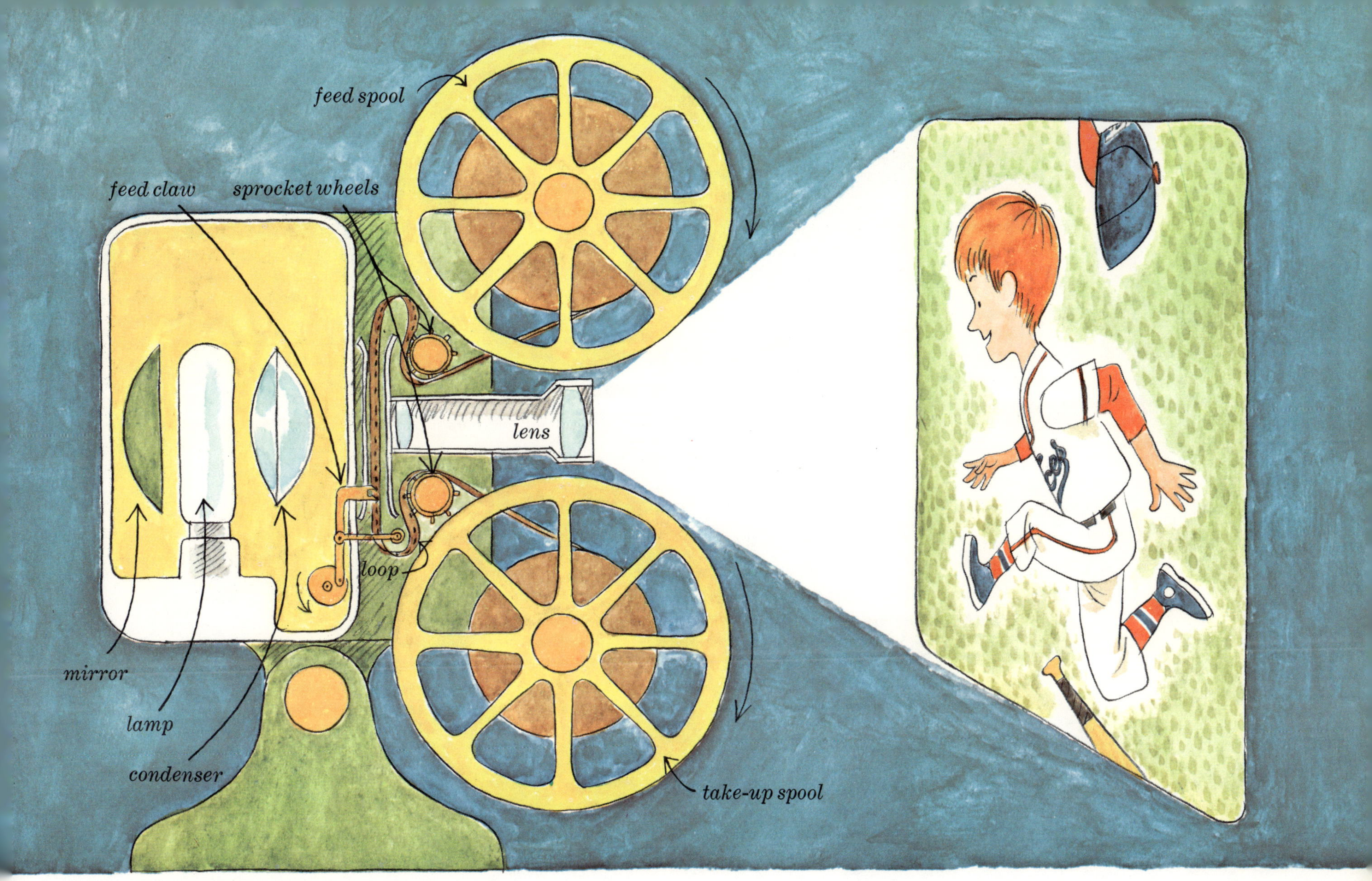

A MOVIE PROJECTOR works just the way a slide projector does, except that it shows a lot of pictures very quickly, one right after the other, so that they appear to be moving. The movie projector has the same kind of mirror, bulb, condenser set-up as the slide projector, and it has a cooling fan, too.

The film goes through the movie projector the same way it went through the camera. A sprocket wheel moves it from the feed spool to the lens, where a claw moves it past the lens one frame at a time. Then the other sprocket wheel rolls the film onto the take-up spool.

When the film is behind the lens, light from the bulb goes through it and projects that frame on the screen. The film must be covered while it moves ahead to the next frame, otherwise the image on the screen would be blurred. So there is a rotary disc shutter which works just like the one in the camera.

Most professional movies have sound. Reading about the tape recorder (page 78) will help you understand one of the ways movie sound is recorded. The movie film has a strip on one side that is coated with iron oxide particles, just like a tape recorder tape. While the movie is being photographed, the actors' speech is recorded on this strip, right next to the pictures. Later, the sound men might add sound effects and music to this sound track. When the movie is shown, an electromagnet coil in the projector picks up the magnetic pattern from the sound track just as the electromagnet coil in the tape recorder picks it up from a tape.

The sound track runs along next to the pictures. It's a sort of tiny tape recording.

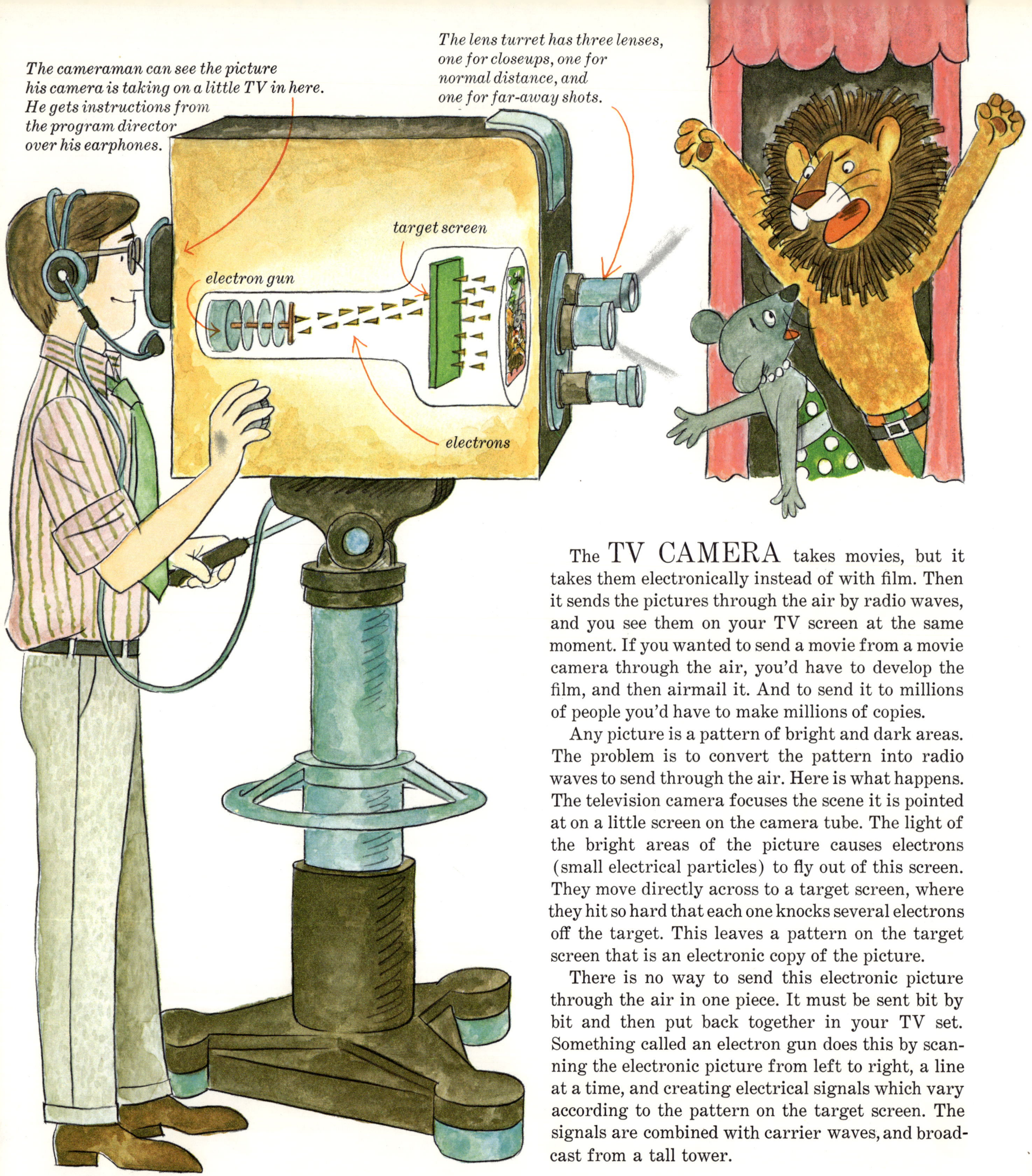

The TV CAMERA takes movies, but it takes them electronically instead of with film. Then it sends the pictures through the air by radio waves, and you see them on your TV screen at the same moment. If you wanted to send a movie from a movie camera through the air, you'd have to develop the film, and then airmail it. And to send it to millions of people you'd have to make millions of copies.

Any picture is a pattern of bright and dark areas. The problem is to convert the pattern into radio waves to send through the air. Here is what happens. The television camera focuses the scene it is pointed at on a little screen on the camera tube. The light of the bright areas of the picture causes electrons (small electrical particles) to fly out of this screen. They move directly across to a target screen, where they hit so hard that each one knocks several electrons off the target. This leaves a pattern on the target screen that is an electronic copy of the picture.

There is no way to send this electronic picture through the air in one piece. It must be sent bit by bit and then put back together in your TV set. Something called an electron gun does this by scanning the electronic picture from left to right, a line at a time, and creating electrical signals which vary according to the pattern on the target screen. The signals are combined with carrier waves, and broadcast from a tall tower.

The TELEVISION SET changes the electrical signals back into pictures and sounds. The antenna receives the broadcast waves, and the tuner selects the wave length of the channel you want. The electrical signals representing the picture are separated from the carrier wave and go to an electron gun at the back of the picture tube.

This electron gun "paints" the picture on the TV screen by shooting the electrons in the signal current at the screen. The screen is coated with a chemical (called a phosphor) that glows for a moment when hit by electrons. The electron gun starts at the top left—the same place the gun in the camera started. It draws a line across the screen from left to right and then another and another, until it recreates all the lines of the picture on the TV screen. This all happens so fast (thirty complete pictures every second) that you see one continuous moving picture. But if you get close to the screen for a moment you can see the separate lines.

Don't worry if you find television hard to understand. It *is* very complicated, and that's why it took a long time for it to be invented. This story might help. Suppose you have a sweater made of black-and-white-spotted yarn. The spots make the special pattern of the sweater. If you want to send this sweater to a friend on the other side of a locked door, you would have to put it through the keyhole. It won't fit all at once, but you can take the end of the yarn where you started knitting, and pass the strand through the keyhole. Now, you start unraveling the sweater, row by row, and your friend starts knitting. As long as he knits the same number of stitches in each row that you did, he'll end up with the same pattern you had. In the TV, the electron guns do the unraveling and the knitting.

RADAR is a system for telling in which direction an object is and how far away it is, even though you can't see it. The large radar antenna sends out beams of radio waves as it sweeps around in a circle. The waves travel in straight lines. If they hit an object, they bounce back to the radar station like an echo bounces back from a cliff. The same antenna picks up the return signal, which goes along wires to the receiver, where it shows up as a mark on the circular radar screen. The location of the mark on the circle shows the direction the antenna was pointing when it picked up the return signal, so it shows the direction of the object.

The distance is known from the time it takes for the radio waves to make the round trip from the antenna to the object and back again, traveling at their constant speed (the speed of light). The electronic circuits do the figuring of the distance and put the mark on the radar screen in the right place. The radar station is in the center of the screen, so the closer to the center the mark is, the closer to the radar station the object is.

This typhoon was photographed by a weather satellite 450 miles above the earth's surface.

SATELLITE. A rocket sends the satellite up into the sky and then into an orbit around the earth. It stays in orbit like a weight on the end of a string does when you whirl it around your head. The weight keeps trying to go off in a straight line, but the string pulling on it keeps it from flying away. In the same way, a satellite is sent whirling around by the rocket, and would go off into space if it were not kept from flying away by the pull of gravity. So it continues to go around in orbit for many years.

A weather satellite has television cameras which take pictures of the earth day and night, showing where the clouds are. It sends the pictures back to earth by radio waves. Weathermen then look at the pictures and can tell what the weather will be like. Satellites are also used to transmit radio, television, and telephone calls over long distances.

Scientists and astronauts can live in a satellite space station, orbiting the earth. They can work there for weeks at a time, doing experiments and finding out more about space.

Oh Wow! the FUTURE. What will they think of next? Perhaps a traveling machine that holds many people and can ride the roads, sail the seas, fly over mountains and, with an attached rocket or two, zip up to the moon for a week-end.

And what will a city of the future be like? It might be enclosed by a huge transparent dome. Inside, the weather would always be pleasant, and the air would always be clean. Rain and melted snow would run off the outside of the dome into huge collecting tanks, so there would always be enough water. The sidewalks might be moving sidewalks, just like escalators running on the level.

What will life in the future be like? You might live in a house that keeps itself clean, and has a recycling unit, into which you put all your cans, bottles, and papers. And what about the stove, refrigerator, and other kitchen appliances—will they change? When you want dinner, you might just press a button, and your food will be thawed, cooked, and delivered to you through a slot, in seconds.

Instead of having a separate television, telephone, and record player, as some people do today, you might have one single machine, a home communications center, that would do all the things they do. This machine might also be able to project slides, movies, and video tapes. While people today have shelves of books, people in this future city might also have shelves of video tape cassettes which could be put into the tape-playing part of the home communications center and watched and heard. These would be talking books with moving pictures.

And, in the future, there will be a new book, or a new kind of book, to explain how it all works.

Blow up a balloon . . .
VOOM
VOOM
let it go . . .
VOOM
. . . and it's like a jet engine.
(See page 28.)

Put a cooking pot into the water.
Even if it's a heavy iron pot, will it sink?
No! It will float like a ship. Why?
(See page 22.)

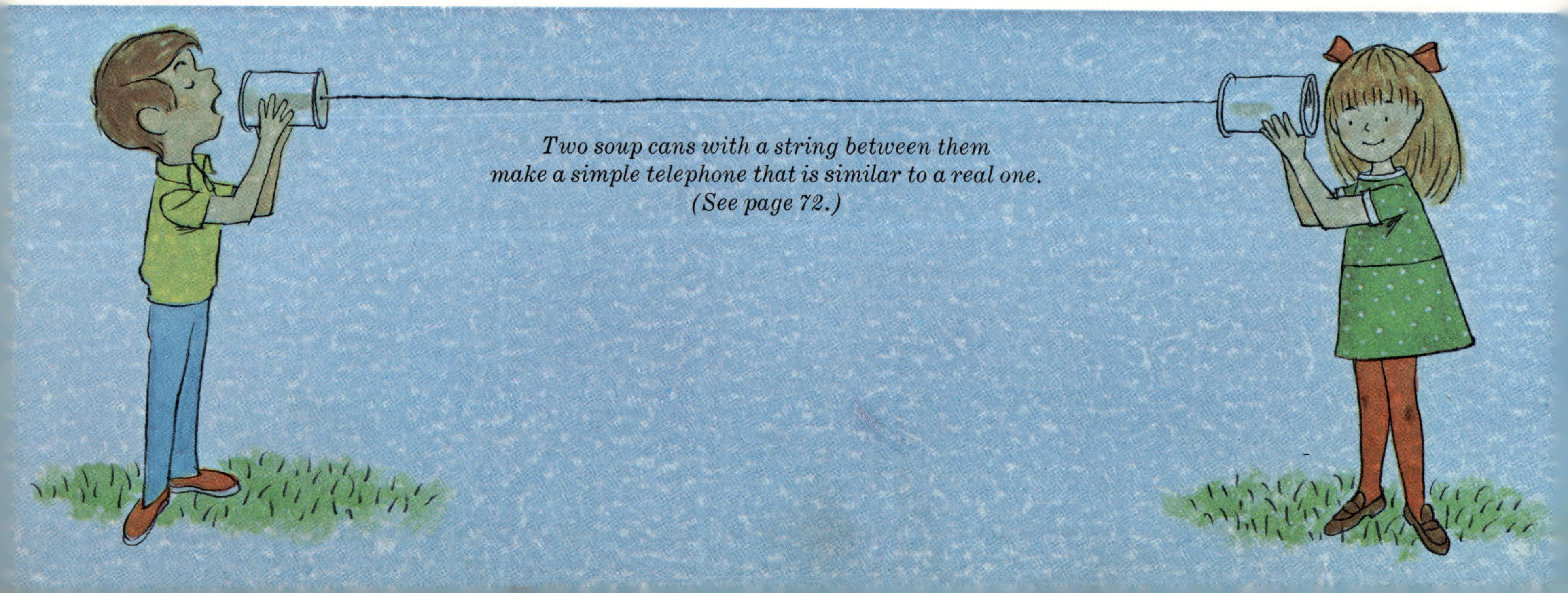
Two soup cans with a string between them
make a simple telephone that is similar to a real one.
(See page 72.)